D1567378

Longevity Records:
Life Spans of Mammals, Birds,
Amphibians, Reptiles, and Fish

Longevity Records: Life Spans of Mammals, Birds, Amphibians, Reptiles, and Fish

James R. Carey and Debra S. Judge

Monographs on Population Aging, 8.

Odense University Press

Longevity Records:
Life Spans of Mammals, Birds,
Amphibians, Reptiles, and Fish

Printed by Special-Trykkeriet Viborg a-s, Denmark
Cover illustration by Allan Waring, Denmark
ISBN 87-7838-539-3
ISSN 0909-119X

Odense University Press
Campusvej 55
DK-5230 Odense M
Phone +45 66 15 79 99 - Fax +45 66 15 81 26
E-mail: Press@forlag.sdu.dk
Internet bookstore: www.oup.dk

Table of Contents

Authors

James R. Carey
Department of Entomology
University of California
Davis, California 95616
and
Center for the Economics and Demography of Aging
University of California
Berkeley, California 94720

Debra S. Judge
Department of Anthropology
University of California
Davis, CA 95616

Preface

A large project such as this cannot be completed without the help of many people. We are grateful to UC Davis undergraduate students Karen Abalos, Harold Amogan, Sophie Betts, Alician Cook, Kim Day, Hoa Giang, Erin Kamei, Sandra Lara, Sarah Mertz, Sakura Nakahara, Phuoc Nguyen, Janel Rodas, Jennifer Schmidt, Nancy Toy, Duc Vo, and Elmer Yee for their help in library research, to Mei Huang for her technical assistance, to Kathy Gruenfelder for supervising student assistants and for initially organizing the data bases, and to Justin Adams for his editorial assistance. Dr. Steven Austad and Dr. Daniel Promislow kindly provided access to their databases. We thank James W. Vaupel for encouraging us to publish the longevity database, Odense University Press for their willingness to allow us to make the data available on the worldwide web, and to Bernard Jeune and Kirsten M. Gauthier for their editorial assistance in getting the book published. This work was supported in part by a Duke University Pilot Project Program grant from the Center for Demographic Studies and by grants from the National Institute on Aging (Grants AG08761-01 and AG14228). Editing and printing were financed by a Danish Research Councils grant (#9502133) to the Aging Research Center, University of Southern Denmark, Odense University.

Introduction

This book contains nearly 4,100 longevity records (reports of highest documented age) for 3,054 vertebrate species/subspecies including 890, 817, 120, 777, and 450 mammals, birds, amphibians, reptiles, and fish, respectively. The number of records exceeds the number of species because for some species we found separate longevity records classified by sex and/or by living conditions (wild or captive) and for others we included two or more record ages.

Longevity records do not represent so-called species-specific maximal ages. First, whereas the life span of an individual animal is unambiguously the interval between its birth and its death, "maximum longevity" is not an appropriate general concept. As Caughley (1977) notes, this is because an animal dies before the age of infinity, not because it cannot pass some bounding age but because the probability of its riding out the ever present risk of death for that long is infinitesimally small. In other words, there is no identifiable age for each species to which some select individuals can survive but none can live beyond. Second, the record age of a species is heavily influenced by the number of individuals observed. That is, the longevity records for species in which the life spans of large numbers of individuals have been observed will be significantly greater than the corresponding figure for a species with the same longevity but represented by a few dozen individuals (Gavrilov and Gavrilova 1991). For the vast majority of longevity records by species, the population at risk and therefore the denominator are completely unknown. Despite some of the analytical and conceptual shortcomings, we believe that the longevity records contained in this book are useful in a number of comparative and disciplinary contexts including demographic, gerontological, ecological and evolutionary. The longevity records provide frames-of-reference for life course analyses for selected species, shed light on the relative longevity of different groups (e.g. mammals vs birds), and help to situate human longevity in the broad context of all living vertebrates.

We organized the book into five sections. The first section introduces the book, describes the data collection methods, describes the major sources of longevity data for each of the vertebrate groups, and presents a brief summary graphic of the longevity data contained in the book. This is followed by a list of selected references on longevity. The subsequent four sections contain the longevity records for each of the four main vertebrate groups. Each section begins with a broad description of the group (e.g. Mammals) followed by a general description for each of the main subgroups (generally Orders, e.g. Artiodactyla), and finally a table with longevity records. These records are organized alphabetically by order, family, and genus-species name.

Data Collection Methods

Major Sources of Longevity Data

Mammals

Major sources for mammalian life spans included the following: (1) Walker's Mammals of the World (Nowak 1991)—this two-volume series for use by professional mammalogists and biologists contains detailed life histories, physical descriptions, and biological information on 1,492 species from 135 families and 21 orders. Walker's texts characterize the world distribution and density of mammals, describe mammalian orders, and present life histories with supplemental black and white photographs. Life span data meeting our criteria was identified for 616 species from 112 families and 21 orders. Wild versus captive species was noted for less than 10% of the animals listed, and sex was specified for roughly one third of the total species. (2) Grzimek's Encyclopedia of Mammals (Parker 1989)—this five-volume series presents life history information in both tables and text for 1,193 species. We identified specific life span information for 544 species from 81 families and 18 orders. Both wild and captive descriptors were noted for less than 15% of the species, and sex was cited for roughly 1% of the total species (e.g., 6 of 544 species reported male or female sex). (3) MacDonald's Encyclopedia of Mammals (MacDonald, 1984)—broad information and illustrations describe life history information on 1,334 species. Life span data meeting our guidelines was identified for 137 species from 49 families and 13 orders. Wild or captive status was noted for roughly half of the animals for which life span data was found. Sex of the record individual was specified for approximately one half of the life span entries. (4) Science and Technology Desk Reference (Bobick and Peffer 1993)—This reference is a compendium of facts and figures put together by Gale Research, Inc. which respond to questions commonly asked of the staff at the Science and Technology Department in the Carnegie Library of Pittsburgh.

Birds

Life span data representing the nearly 10,000 species, 29 orders, and 187 families of birds as depicted by Gill (1995) was extracted from four primary sources. The life span database on birds includes species from 27 orders and 185 families. The following key sources are described in order of their relative contribution to the information identified on the life span of birds. (1) Handbook of Birds of the World (del Hoyo et al. (eds.) 1992a,b)—these two volumes of handbooks produced by the International Council for Bird Preservation (ICBP) provide life history and conservation information on birds obtained via support from ornithologists and amateurs from around the world. Life histories organized by family texts characterize species' breeding, habitat, and appearance, and are supplemented by bird plates and distribution maps for 100 species. Life span information was found for roughly 100 species from 11 orders. (2) Birds of North America (Poole, Stettenheim and Gill (eds.) 1994)—this continuing project, sponsored by the American Ornithologist's Union and the Academy of Natural Sciences of Philadelphia, will grow to reach a proposed goal of 700 species' profiles at the rate of 40-80 profiles per year which will be included in the volumes of "Birds of North America." Single species' profiles are published individually throughout the year. The profiles are written for conservationists, ornithologists, and naturalists by specialists on each specific taxon, and life span data is reported for almost every bird species described in volumes one through five. (3) Animal Life Spans (Flower 1938)—Major Stanley S. Flower, a former curator of the London Zoological Gardens, interviewed and recorded factual life span information obtained from private collectors and zoos of Europe and North America for more than 250 species. The Proceedings of the General Meetings for the Scientific Business of the Zoological Society of London consolidate and publish papers presented at the Society's meetings on an annual basis. Flower's records met our life span data guidelines for over 100 species from 15 orders. (4) Clapp, Klimkiewicz et al. (1982; 1983)—these references report life span records for more than 200 species based on bird-banding records collected via mark-and-retrieval of birds living in the wild. Their articles titled "Longevity Records of North American Birds: Gaviidae through Alcidae," and "Longevity Records of North American Birds: Columbidae through Paridae," have been published in The Journal of Ornithology (formerly Bird-Banding) which is produced by the Northeastern Bird-Banding Association for use by ornithologists and other professionals. The authors report maximum ages of birds reported to have been killed, found dead, or recaptured alive. Sex was noted for approximately 20% of the entries. (5) Klimkiewicz and Futcher (1987, 1989)—the authors add to the 1982 and 1983 compilations of longevities of marked birds in their article, "Longevity Records of North American Birds: Supplement I," which was also published in the Journal of Field Ornithology.

Amphibians and Reptiles

The following sources together contributed more than 75% of the species for which we identified life span information. (1) Longevity of Reptiles and Amphibians in North American Collections (Bowler 1975)—this reference reviews existing knowledge of amphibian and reptile life spans in captivity. As published for the Society for the Study of Amphibians and Reptiles and as written from the perspective of the Curator of Reptiles from the Zoological Society of Philadelphia, Bowler examined ages reported from institutions and from individuals. (2) Goin, Goin, and Zug (1978)—the authors intended the text as a foundation for understanding the nature of herpetology and as an introduction to herpetology research. With a focus on natural history, the authors describe both growth and life span parameters providing more than 50 species' life spans. (3) Animal Life Spans (Flower 1925a-e, 1936, 1937). Stanley S. Flower published records of amphibian and reptile life spans in captivity recorded predominantly in zoo records, but also through reports of private collectors. Flower published life span data through the Proceedings of the Zoological Society of London. (4) The Science and Technology Desk Reference (Bobick and Peffer 1993)—this reference is explained above.

Fishes

The majority of life spans for fishes came from six key sources that are described below. Primary resources are two articles written by researchers studying fish, two reference guides, and two reports on fish longevity. Key sources include the following: (1) Beverton and Holt (1959)—Beverton and Holt of the Ministry of Agriculture, Fisheries and Food, Fisheries Laboratory of Rome examined the life span and force of natural mortality in fish populations. Their work overviews growth patterns of captured fish for the purpose of conservation. Species name, common name, location, parameters of a growth equation, and life span are available for up to 90 species. (2) Altman and Dittmer (1991)—the Committee on Biological

Handbooks in Washington has prepared biology handbooks for the Federation for Experimental Biology since 1964. The Biology Data Book reports life spans of 233 vertebrate and 92 invertebrate species held in captivity. With 124 different sources of life span data reported in this data book, authors selected Recorded Maximum Life Span in years and months, and reported age determination method (e.g., by growth layers, mark-and-recapture, unknown age, or specimen still alive at the time of age-reporting) information for presentation in their tables. Life spans for 37 were extracted from Volume I of the Second Edition of the Data Book for species in captivity. Life spans of 269 species were extracted from the Growth Edition of the Data Book for species in captivity or in nature. (3) Fishes of the Sacramento-San Joaquin Estuary and the Adjacent Waters, California: A Guide to the Early Life Histories (Wang 1986)—this reference provides information in a technical report on California fish ecology and conservation. We identified around 50 species' life spans from this source. (4) The Science and Technology Desk Reference (Bobick and Peffer 1993)—this is a compendium of facts and figures which respond to the wealth of questions commonly asked of the staff at the Science and Technology Department at the Carnegie Library of Pittsburgh. (5) Distribution and Abundance of Fishes and Invertebrates in West Coast Estuaries, Vol. II: Species Life History Summaries (Emmett 1991)—this was written for the Strategic Environmental Assessments Division in Rockville, MD to establish the nature and status of fish populations along the West Coast. We identified approximately 30 life spans of fishes from this source. Additional sources of life span information were identified in journals such as Ecology, The Journal of Animal Ecology, California Fish and Game, Journal of the Fisheries Research Board, and Contributions to Canadian Biology and Fisheries.

After the major sources were surveyed and values recorded, a search of primary literature using AGRICOLA, BIOSIS© and Current Contents© databases was undertaken. Key words or title words employed in the search include life span, life span, longevity, maximum age and life table. The data reported in this book originate in approximately 700 published sources.

Exclusion Criteria

Longevity data were excluded based on the following criteria: (1) Uncertainty of the estimate—longevity information failed to list a specific age (e.g., "several years," "a few years," "following first mating," "about x-years," "same as other animals within the group," etc.). (2) Average life span—life span data cited an average rather than a maximum age. (3) Unspecified species—life span was listed for all members of a group larger than the genus (e.g., a subfamily or order); (4) Non-factual information—the life span information was formulated based on myths of extraordinarily long-lived species which are undocumented.

Data Characteristics

For each record life span that passed the exclusion criteria, the following data was ascertained and included in this database: (1) Sex—when specified the sex was noted as m (= male) or f (= female). Unspecified sex is noted as x (= not noted in reference). (2) Captive/Wild—for each record life span it is noted whether the individual animal was captive or wild (when such was reported) to reflect the biases of environmental factors and aging techniques. (3) Age—life spans for birds, amphibians and reptiles, and mammals are recorded in years with months interpreted as a proportion of the full 12-month year to one decimal place (e.g. 10 years 4 months is recorded as 10.3 yrs). (4) Captive Age—if the age of an adult when captured is unknown, the time held in captivity subsequent to capture is reported as the life span. This life span is recorded as a minimum, denoted by a (+) following the species' common name, since the duration of life prior to capture has not been recorded. (5) Maximum Age—if an age range is given, then the highest number is taken as maximum life span and the lower number is discounted. (6) Minimum Age—if the reported age is 'at least' x-years, then the life span is a minimum age and is denoted by a (+) following the species' common name. (7) Each record is followed by specific reference(s) where that observation is published. The full references for each major vertebrate subgroup are listed immediately after each table of records.

Summary

A graphical summary of the longevity data for each of the four broad vertebrate groups is presented in Fig. 1. This figure captures a number of different patterns regarding longevity and vertebrates that other researchers have identified and can be summarized as follows (see Selected References). First, longevity is positively correlated with body size between orders (e.g. the smaller rodents are shorter lived than the larger cetaceans) though not necessarily within orders (e.g. longevity not correlated with body size in the pinnipeds; i.e. seals and walruses). Second, animals that fly (i.e. birds and bats), are armored (turtles; armadillos) or live underground (moles; mole rats) tend to live longer than is predicted from body size alone. Third, life spans differ by a factor of over 50 in mammals, herps and fish and by over 15-

fold in birds; body size, metabolic rate, brain size all being positively correlated with life span. Fourth, primates are long-lived mammals, the great apes (i.e. gorillas; chimpanzees) are long lived primates, and humans are extraordinarily long-lived great apes; human longevity exceeds nearly all other species both relatively and absolutely.

Figure Legend

Fig. 1. Summary chart of all longevity data contained in the book for mammals, birds, reptiles/amphibians and fishes. Subgroups such as Orders (mammals) or Superfamilies (birds) within each section were ordered by mean longevity within the group. Statistics including mean, standard deviation and ranges were determined using the maximal age for each species regardless of sex or whether this age was determined from individuals in the wild or from laboratory/zoo records. The objective of this chart is to provide a visual summary of broad trends within and between groups. The rank-ordering (top-to-bottom) by mean longevities for each of the five groups are as follows (see introductory sections for common names):

Mammals

Proboscidae, Cetacea, Perrisodactyla, Pinnepeds, Monotremata, Primates, Edentata, Artiodactyla, Carnivora, Chiroptera, Hyracoidea, Marsupialia, Lagomorpha, Rodentia, Macrocelidae and Insectivora.

Birds

Struthioniformes, Psittaciformes, Coraciiformes, Sphenisciformes, Procellariformes, Gaviformes, Pelecaniformes, Casuariformes, Anseriformes, Ciconiiformes, Gruiformes, Falconiformes, Charadriiformes, Strigiformes, Columbiformes, Piciformes, Galliformes, Passeriformes, Podiformes, Apodiformes, Trochiliformes, Caprimulgiformes, and Cuculiiformes.

Herps

Crocodilians, Testidines, Squmata, Gymnophiona, Anurans, and Lizards.

Fish

Ancipenseriformes, Anguilliformes, Pleuronectiformes, Scorpaeinformes, Gadiformes, Cartilaginous, Siluriformes, Cypriniformes, Perciformes, Salmoniformes, Clupeiformes, Atheriniformes, and Gasterosteiformes.

Figure 1.

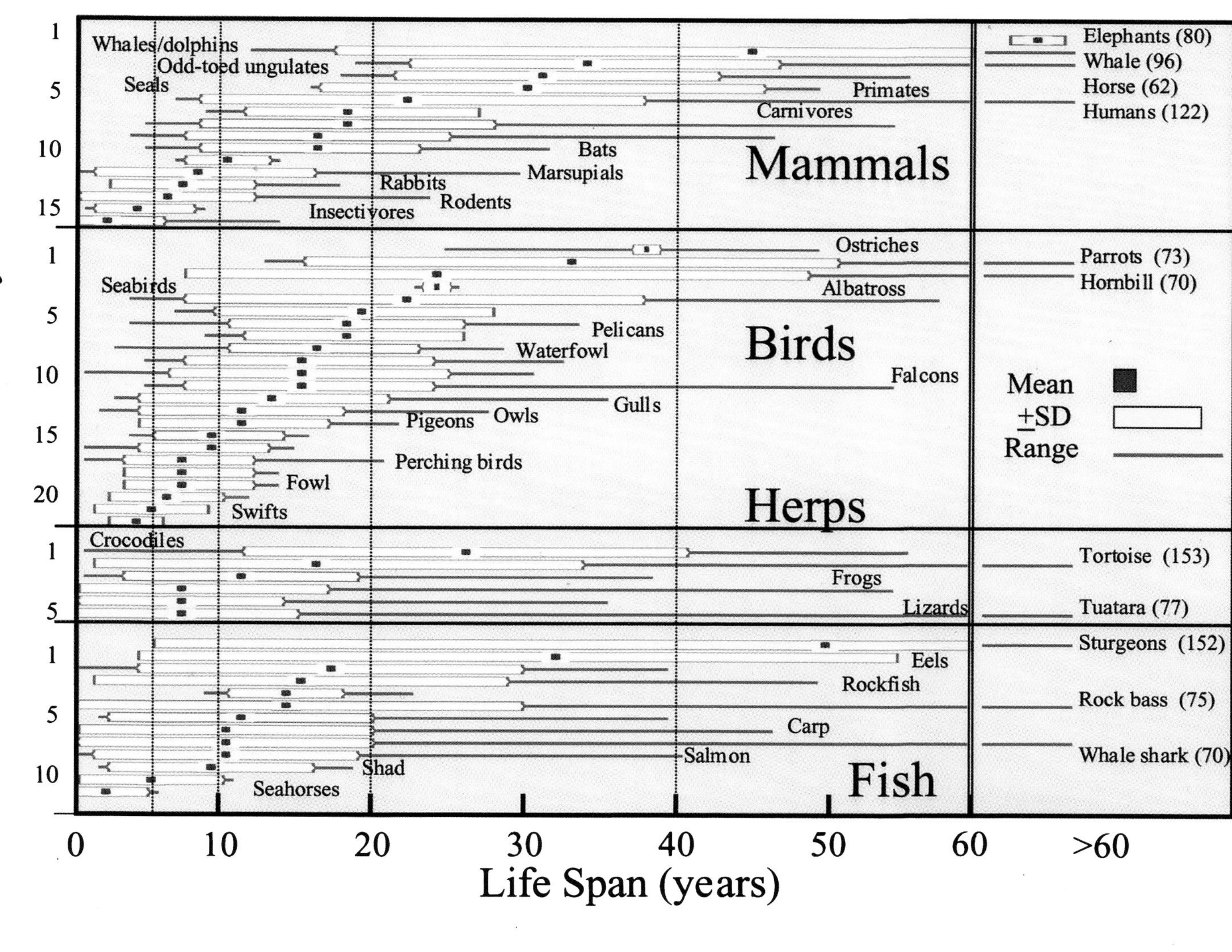
Order/Family
Life Span (years)
0
10
20
30
40
50
60
>60
Whales/dolphins
Odd-toed ungulates
Seals
Primates
Carnivores
Bats
Marsupials
Rabbits
Rodents
Insectivores
Mammals
Elephants (80)
Whale (96)
Horse (62)
Humans (122)
Ostriches
Seabirds
Albatross
Pelicans
Waterfowl
Birds
Falcons
Gulls
Pigeons
Owls
Perching birds
Fowl
Swifts
Parrots (73)
Hornbill (70)
Mean
±SD
Range
Herps
Crocodiles
Frogs
Lizards
Tortoise (153)
Tuatara (77)
Eels
Rockfish
Carp
Salmon
Shad
Seahorses
Fish
Sturgeons (152)
Rock bass (75)
Whale shark (70)

References Cited

Altman, P. L. and D. S. Dittmer, 1972. Biology Data Book. Bethesda, Federation of American Societies for Experimental Biology.

Austad, S. N. and K. E. Fischer, 1991. "Mammalian aging, metabolism, and ecology: evidence from the bats and marsupials." *Journal of Gerontology* 46(2): 47-53.

Austad, S. N. and K. E. Fischer, 1992. "Primate longevity: its place in the mammalian scheme." *American Journal of Primatology* 28(4): 251-261.

Beverton, R. J. H. and S. J. Holt ,1959. A review of the lifespans & mortality rates of fish in nature and their relationship to growth and other physiological characteristics. *Ciba Foundation Colloquia on Ageing* 5:142-180.

Bobick, J. E. and M. Peffer, Eds., 1993. *Science and Technology Desk Reference.* Washington, D.C., Gale Research Inc.

Bowler, J. K., 1975. "Longevity of reptiles and amphibians in North American collections as of 1 November, 1975." *Society for the Study of Amphibians and Reptiles, Miscellaneous Publications, Herpetological Circular No.6*(6): 1-32.

Caro, T. M., D. W. Sellen, et al., 1995. "Termination of reproduction in nonhuman and female primates." *International Journal of Primatology* 16: 205-220.

Caughley, G., 1977. *Analysis of Vertebrate Populations*. Chichester, Australia, John Wiley & Sons.

Charnov, E. L., 1990. "On evolution of age of maturity and the adult lifespan." *Journal Evolutionary Biology* 3: 139-144.

Clapp, R. B., M. K. Klimkiewicz, et al., 1982. "Longevity records of North American birds: Gaviidae through Alcidae." *Journal of Field Ornithology* 53: 81-208.

Clapp, R. B., N. K. Klimkiewicz, et al., 1983. "Longevity records of North American birds: Columbidae through Paridae." *Journal of Field Ornithology* 54: 123-137.

Comfort, A., 1961. "The life span of animals." *Scientific American* 205: 108-119.

Comfort, A., 1964. *Ageing: The Biology of Senescence*. London, Routledge and Kegan Paul.

Cutler, R. G., 1976. "Evolution of longevity in primates." *Journal of Human Evolution* 5: 169-202.

delHoyo, J., A. Elliot, et al., Eds., 1992. *Handbook of Birds of the World, Vol. 1.* Barcelona, Lynx Edicions.

delHoyo, J., A. Elliott, et al., Eds., 1992. *Handbook of Birds of the World, Vol. 2.* Barcelona, Lynx Edicions.

Economos, A. C., 1980. "Brain-life span conjecture: a reevaluation of the evidence." *Gerontology* 26: 82-89.

Economos, A. C., 1980. "Taxonomic differences in the mammalian life span-body weight relationship and the problem of brain weight." *Gerontology* 26: 90-98.

Emmett, R. L. ,1991. Distribution and Abundance of Fishes and Invertebrates in West Coast Estuaries. II: Species Life History Summaries. Rockville, MD, U.S. Dept. of Commerce; National Oceanic and Atmospheric Administration, National Ocean Service: 329.

Finch, C. E., 1990. *Longevity, Senescence and the Genome.* Chicago, University of Chicago Press.

Finch, C. E. and M. C. Pike 1996. "Maximum life span predictions from the Gompertz mortality model." *Journal of Gerontology* 51A(3): B183-B194.

Flower, M. S. S., 1925. Contributions to our knowledge of the duration of life in vertebrate animals - III. Reptiles. *Proceedings of the Zoological Society of London.* London, Printed for the Society: 911-981.

Flower, M. S. S., 1925. Contributions to our knowledge of the duration of life in vertebrate animals - I. Fishes. *Proceedings of the Zoological Society of London.* London, Printed for the Society: 247-267.

Flower, M. S. S., 1925. Contributions to our knowledge of the duration of life in vertebrate animals - II. Batrachians. *Proceedings of the Zoological Society of London.* London, Printed for the Society: 269-289.

Flower, M. S. S., 1925. Contributions to our knowledge of the duration of life in vertebrate animals - IV. Birds. *Proceedings of the Zoological Society of London.* London: 1365.

Flower, M. S. S., 1925. Contributions to our knowledge of the duration of life in vertebrate animals - V. Mammals. *Proceedings of the Zoological Society of London.* London: 145.

Flower, M. S. S., 1935. Further notes on the duration of life in animals. - I. Fishes: as determined by otolith and scale - readings and direct observations on living animals. *Proceedings of the Zoological Society of London*: 265.

Flower, M. S. S., 1936. Further notes on the duration of life in animals - II. Amphibians. *Proceedings of the Zoological Society of London.* London, Printed for the Society: 369-394.

Flower, M. S. S., 1937. Further notes on the duration of life in animals - III. Reptiles. *Prooceedings of the Zoological Society of London*. London, Printed for the Society: 1-39.

Flower, M. S. S., 1938. The duration of life in animals - IV. Birds: special notes by orders and families. *Proceedings of the Zoological Society of London*. London, Printed for the Society: 195-235.

Flower, M. S. S., 1948. Further notes on the duration of life in mammals - V. The alleged and actual ages to which elephants live. *Journal of Zoology: Proceedings of the Zoological Society of London*. 117: 680.

Gavrilov, L. and N. Gavrilova, 1991. *The Biology of Life Span*. Chur, Switzerland, Harwood Academic Publishers.

Gill, F. B., 1995. *Ornithology*. New York, W. H. Freeman and Company.

Goin, C. J., O. B. Goin, et al., 1978. *Introduction to Herpetology*. San Francisco, W. H. Freeman and Company.

Hakeem, A., R. Sandoval, M. Jones, and J. Allman, 1996. Brain and lIfe span in primates. *Handbook of the Psychology of Aging*. J. Birren, Academic Press: 78-104.

Hammer, M. L. A. and R. Foley, 1996. "Longevity, life history and allometry: how long did hominids live?" *Human Evolution* 11: 61-66.

Hayflick, L., 1987. Origins of longevity. *Modern Biological Theories of Aging*. H. R. Warner, R. L. Sprott, R. N. Butler and E. L. Schneider. New York, Raven Press.

Hirsch, H. R., 1994. "Can an improved environmental cause maximum lifespan to decrease? comments on lifespan criteria and longitudinal Gompertzian analysis." *Experimental Gerontology* 29: 119-137.

Jeune, B. and J. W. Vaupel, Eds., 1995. *Exceptional Longevity: From Prehistory to the Present*. Monographs on Population Aging. Odense, Denmark, Odense University Press.

Jeune, B. and J. W. Vaupel, Eds., 1999. *Validation of Exceptional Longevity*. Monographs on Population Aging. Odense, Denmark, Odense University Press.

Kirkwood, T. B. L., 1992. "Comparative life spans of species: why do species have the life spans they do?" *American Journal of Clinical Nutrition* 55: 1191S-1195S.

Klimkiewicz, M. K., R. B. Clapp, et al., 1983. "Longevity records of North American birds: Remizidae through Parulinae." *Journal of Field Ornithology* 54: 287-294.

Klimkiewicz, M. K. and A. C. Futcher, 1989. "Longevity records of North American birds: supplement 1." *Journal of Field Ornithology* 60: 469-494.

Klimkiewicz, M. K. and A. G. Futcher, 1987. "Longevity records of North American birds: Coerbinae through Estrildidae." *Journal of Field Ornithology* 58: 318-333.

MacDonald, D., Ed. 1984. *The Encyclopedia of Mammals*. New York, Facts on File, Inc.

Nowak, R. M., 1991. *Walker's Mammals of the World, Vol. I.* Baltimore, Johns Hopkins University Press.

Nowak, R. M., 1991. *Walker's Mammals of the World. Vol. II.* Baltimore, Johns Hopkins University Press.

Olshansky, S. J., B. A. Carnes, et al., 1990. "In search of Methuselah: estimating the upper limits to human longevity." *Science* 250: 634-639.

Parker, S. P., Ed. 1989. *Grzimek's Encyclopedia of Animals*. New York, McGraw-Hill Publishing Company.

Partridge, L. and P. H. Harvey, 1988. "The ecological context of life history evolution." *Science* 241: 1449-1455.

Patronek, G. J., D. J. Waters, et al., 1997. "Comparative longevity of pet dogs and humans: implications for gerontology research." *Journal of Gerontology* 52A: B171-B178.

Poole, A., P. Stettenheim, et al., Eds., 1992. *The Birds of North America*. Washington, DC, American Ornithologist's Union, Philadelphia Academy of Natural Sciences.

Promislow, D. E. L., 1993. "On size and survival: progress and pitfalls in the allometry of life span." *Journal of Gerontology* 48: B115-B123.

Promislow, D. E. L. and P. H. Harvey, 1990. "Living fast and dying young: a comparative analysis of life-history variation among mammals." *Journal of Zoological Society of London* 220: 417-437.

Prothero, J. and K. D. Jurgens, 1987. Scaling of maximal lifespan in mammals: a review. *Evolution of Longevity in Animals*. A. D. Woodhead and K. H. Thompson. New York, Plenum Press: 49-74.

Ricklefs, R. E., 1998. "Evolutionary theories of aging: confirmation of a fundamental prediction, with implications for the genetic basis and evolution of life span." *American Naturalist* 152: 24-44.

Sacher, G. A., 1975. Maturation and longevity in relation to cranial capacity in hominid evolution. *Primate Functional Morphology and Evolution*. R. H. Tuttle. The Hague, Mouton: 417-442.

Sacher, G. A., 1977. Life table modification and life prolongation. *Handbook of the Biology of Aging*. C. E. Finch, L. Hayflick, H. Brody, I. Rossman and F. M. Sinex. New York, Van Nostrand Reinhold Co.: 771.

Sacher, G. A., 1978. "Longevity and aging in vertebrate evolution." *Bioscience* 28: 497-501.

Sacher, G. A., 1980. "The constitutional basis for longevity in the Cetacea: do the whales and the terrestrial mammals obey the same laws?" *Report of the International Whaling Commission (Special Issue 3)*(3): 209-213.

Sacher, G. A., 1980. "Theory in gerontology." *Annual Review Gerontology and Geriatrics* 1: 3-25.

Sacher, G. A., 1982. "Evolutionary theory in gerontology." *Perspectives in Biology and Medicine* 25: 339-353.

Smith, D. W. E., 1997. "Centenarians: human longevity outliers." *The Gerontologist* 37: 200-207.

Vaupel, J. W., J. R. Carey, et al., 1998. "Biodemographic trajectories of longevity." *Science* 280: 855-860.

Wachter, K., 1997. Between Zeus and the salmon. Introducing the biodemography of longevity. *Biodemography of Aging*. K. Wachter and C. Finch. Washington, DC, National Academy Press.

Wang, J. C. S. ,1986. Fishes of the Sacramento-San Joaquin estuary and adjacent waters, California: A guide to the early life histories. Sacramento, Department of Water Resources.

West, G. B., J. H. Brown, et al., 1999. "The fourth dimension of life: fractal geometry and allometric scaling of organisms." *Science* 284: 1677-1679.

Wilmoth, J. R. and H. Lundstrom, 1996. "Extreme longevity in five countries." *European Journal of Population* 12: 63-93.

Mammals

Roughly 4,500 species in 19 orders make up the mammals (Nowak 1991). The largest order of mammals is the the Rodentia with over 1840 species, while bats (Chiroptera) are the second largest order with almost 1000 species. Mammals are characterized by homeothermy, hair, relatively large brains, and mammary glands for postnatal maternal nutrition of young.

The primitive Protheria include the Monotremata (the platypus and echidna; McKenna 1975), while the Theria includes all other mammals (Janson and Pope, 1989). The Theria may be further subdivided as the Metatheria (the pouched mammals or Marsupialia) and the Eutheria (placental mammals). Each mammalian order is briefly described with respect to general ecology, life history characteristics, and distribution.

MONOTREMES (echidna and duck-billed platypus)

Native to Australia and New Guinea, monotremes are the only egg-laying mammalian order. The two Monotremata families are the Ornithorhynidae (Duck-billed Platypuses) and Tachyglossidae (Spiny Anteaters). Spiny anteaters are toothless, spine-covered animals with strong clawed feet for digging and a long sticky tongue. They live in solitarily in crevices and burrows, can torpor, and may survive for a month without insects, worms, or other food (Nowak 1991). Broods of one or three spiny anteater young are born and receive extended maternal care.

Platypus live in burrows near lakes and streams where they feed on small invertebrates and fish (Grant and Carrick 1978). Platypus have short limbs with webbed feet, and a broad, flat tail. Males and females pair-bond seasonally and lay clutches of one to three eggs in a nest (Nowak 1991).

MARSUPIALS (opossums, bandicoots and kangaroos)

The marsupials are pouched mammals native to southern North American and South Americancontinents (70 species) as well as Australia (170 species) (Janson and Pope, 1989). The 280 species of marsupials include the American opossums (Didelphidae, 77 species), carnivorous marsupials (Dasyuridae, 58 species), bandicoots (Peramelidae, 21 species), striped opossums (Petauridae, 23 species), phalangers (Phalangeridae, 20 species), and the kangaroos and wallabies (Macropodidae, 56 species)as well as ten other smaller families. Marsupial body size and diet vary widely and many are nocturnal. Marsupial females have a reproductive system of dual vagina and uteri. Females bear altricial young which are protected in the marsupium (with the exception of some smaller marsupials) and, in some species, young of subsequent litters developing simultaneously though at different stages.

INSECTIVORA (shrews, tenrecs, hedghogs and moles)

The Insectivores are the third largest order of mammals, comprised of almost 400 species in 7 families. The Gymnures and Hedghogs (Erinaceidae, 20 species), Solenodons (Solenodontidae, 2 species), Extinct West Indian Shrews (Nesophontidae, 6 species), Tenrecs and Madagascar "Hedgehogs" (Tenrecidae, 23 species), Golden Moles (Chrysochloridae, 18 species), Shrews (Soricidae, 289 species), Shrew-moles, and Desmans (Talpidae, 32 species) are small, clawed, animals with an elongated narrow snouts.

Most insectivores are nocturnal with diets consisting of grubs, insects, snails, and small vertebrates. Insectivores pervade numerous habitats and are highly active. They seek shelter in enclosures, forests, and foliage. Insectivores have a gestation period of one or two months, variable litter size, and an extended period of parental care (Nowak 1991). Most insectivores live for fewer than five years. Hedgehogs and Solenodons, however, may live into their teens. Solenodons are shrew-like animals with toxic saliva potent enough to immobilize larger prey such as small birds, frogs, and lizards.

MACROSCELIDEA (elephant shrews)

Fifteen species of elephant shrews are native to African forests, grasslands, and thickets. They are small (95-315 mm. head and body length) with large eyes and ears, a long, narrow snout that is not retractile but can be moved in a circular motion, and hind limbs adapted for jumping. Elephant shrews are generally solitary insectivores that produce small broods of precocial young after two months of gestation. Lactation is short and sexual maturity is achieved as early as five weeks (Nowak 1991). Some species of genus Elephantulus reside in colonies and are known to communicate through chirps and tapping sounds made using the hind legs (Nowak 1991).

DERMOPTERA (flying lemurs)

There are two species of Dermoptera in one family native to Southeast Asian mountain or lowland forests. Flying Lemurs are 340-420 mm. body length, with long limbs and tail, large eyes, short ears and nocturnal habits. When outstretched, membranes running from the neck to the front feet, back feet, and end of the tail allow flying lemurs to glide between branches and to feed on flowers and fruit high in tree tops or on outermost tree limbs that are out of the reach of most animals. Activity is reduced during the day when Flying Lemurs hang upside down by their claws. Gestation is 60 days, and only one offspring is born per litter. Young receive extended maternal care and several flying lemurs may live together (Nowak 1991).

CHIROPTERA (bats)

More than 900 species of bats live in most tropical and temperate climes. The suborder Megachiroptera (fruit bats) are tropical and homeothermic. The Microchiroptera are tropical and temperate, commonly heterothermic and insectivorous -- though some eat fruit and flowers and a few species prey on vertebrates. Bats are the only mammals engaging in true flight. Bats are found within the limits of tree growth, and find sanctuary in caves and other small crevices in buildings or trees. Temperate zone bats generally mate in the fall and store sperm through hibernation then produce only one offspring per litter in the spring.

Bats are long-lived for their small body size. Maximum life spans for most species exceed 10 years, while those of others are reported to extend into the thirties.

SCANDENTIA (tree shrews)

Scandentia includes one family (Tupaiidae) of Tree Shrews with five species native to Eastern Asia and Southwest China forests. Difficult to classify, this order is now related most closely to Primates, Chiroptera and Dermoptera. Resembling squirrels with long snouts, Tree Shrews are mostly diurnal, small (100-220 mm. head and body length), bushy-tailed, and feed on insects, fruit, and worms. Most Tree Shrews are adept runners, jumpers, and/or tree-climbers. Communication occurs in snarls and other vocalizations. Shrews appear to be solitary and polygynous; litters of 1 to 3 young are nursed about 5 to 6 weeks and are sexually mature at 3 months (Nowak 1991).

PRIMATES (Prosimians, monkeys and apes)

Primates are generally pan-tropical with the exception of humans who live worldwide (Nowak 1991). The suborder Prosimii includes 6 families (tarsiers, lemurs, indris, aye ayes, galagos, pottos, and lorises), and the Anthropoidea including three New World families (Cebidae, Callitrichidae and Callimiconidae) and three Old World families (Cercopithecidae, Pongidae, Hominidae). Anthropoid primates are diurnal, highly dextrous and have fine manipulative skills (Janson and Pope, 1989) as well as the highest brain to body size ratio. Primates range in size from the mouse lemurs (50 g) to gorillas (200 kg.). Primate faces are flat with a short nose and jaw, forward facing eyes, and teeth designed for grinding. The majority of primates are polygynous with annual production of single young and extended maternal (and sometimes paternal) care.

XENARTHRA (armadillos, anteaters and sloths)

Approximately 30 species of Xenarthra include the armadillos native to the Western Hemisphere (Dasypodidae, 20 species), anteaters native to Central and South America (Myrmecophagidae, 4 species), and sloths (Bradypodidae, 5 species). Head and body lengths range from 125 to 1200 mm (Nowak 1991). Armadillos have bony plates, cylindrical teeth for insect consumption, and digging feet for rooting. Anteaters have a long snout sticky tongue and a diet of ants and termites. Socially, most Xenarthra are solitary but some form loose groups (Nowak 1991). Most sloths bear young singly as do most anteaters, while armadillos bear litters of one to three genetically identical young.

PHILODOTA (spiny anteaters)

Philodota includes one family with seven species of Spiny Anteaters native to parts of Asia, Indonesia, and Africa. They range in head to rump length from 300 to 880 millimeters with scales, short limbs, a cone-shaped skull, and a long tail (Nowak 1991). Like the Xenarthra anteaters, Pholidotes have a stomach that grinds the exoskeletons of insects for more complete digestion. Spiny Anteaters roam in forests, brush, and savanna habitats by night, and remain in burrows or tree crevices by day. They are toothless insectivores who pick up ants, termites, and other foodstuffs using their sticky tongues. When attacked, the Spiny Anteater curls into a ball and may roll down a hill to escape predation, or may eject foul-smelling liquid though more generally they lead slow-moving solitary lives of low aggression (Nowak 1991). Litters of one or two young are born and receive extended maternal care.

LAGOMORPHA (rabbits and hares)

The Lagomorpha consist of two families, the Laporidae (about 50 species of hares and rabbits) and the Ochotonidae (14 species of pikas), native to most countries with the exceptions of New Zealand and Australia. They are characterized by long ears, short fur, an upturned tail, strong hind legs for jumping, and a high reproductive rate.

Rabbits generally live in lowland areas and hills where they build underground burrows and congregate in groups often exceeding 100 members (Janson and Pope, 1989). Their food is double-processed by eating plant food, passing this through the system, and re-consuming the excreted pellets. Except during the mating season, hares are solitary. While rabbits dig underground burrows, hares rest in depressions in the grass. Pikas are found in North America and Asia where they live in a wide range of habitats. Litter size in the Lagomorpha generally ranges from 1-5, although some species bear as many as 15 per litter.

RODENTIA (squirrels, mice, marmots and beaver)

Rodentia is the largest mammalian order and includes the suborders Sciuromorpha (7 families of squirrel-like rodents; squirrels, marmots, gophers, and beavers), Myomorpha (1000 species in 9 families of mouse-like rodents), and Hystricomorpha (16 families of porcupine-like rodents; porcupines, guinea pigs, and coypus). Rodents are pandemic and diverse, including over 1800 species in a wide variety of habitats. Rodents have sharp incisors, grinding teeth in the back, and, in some species, cheek pouches. Though their diets are varied, seeds and vegetative plant parts are predominant.

CETACEA (whales and dolphins)

The 79 species of Cetacea are predominantly marine; however, there are riverine and lacustrine dolphins. Whales may be divided into the Odontoceti (7 families of toothed whales), and three families of the Mysticeti (the baleen whales). Cetacea are characterized by a streamlined body, broad flippers, and blubber insulation. Nostrils at the top of the head are used for breathing, with a single or double blow-hole in toothed and baleen whales, respectively. In deep dives, oxygen is directed to the brain and nerves, and some muscles have evolved to function anaerobically. Cetaceans have an absolute brain size larger than that of humans and like humans the surface is highly convoluted (Nowak 1991).

Cetacean olfaction and vision is limited, but touch and hearing are acute. Whales and dolphins communicate through clicks and whistles. Some Odontoceti have the capacity of echolocation for hunting and navigation (Janson and Pope, 1989). The baleen whales have sieve-like plates that filter crustaceans, zooplankton and fish. Toothed whales (Odontoceti) have cone-like teeth for grasping fish, squid, and cuttlefish which are generally swallowed whole; however, Orca may feed more opportunistically on warm-blooded marine life.

Cetacean young mature slowly after a long gestation period (9-17mos). Baleen whales mature at 12 years and produce young only once every two years (Janson and Pope, 1989). Calves receive extended maternal care and in some species there is evidence of allomaternal care (Whitehead 1996).

CARNIVORA (weasels, skunks, otters, dogs and cats)

There are 240 species in 7 families of Carnivores. They live throughout the world and are characterized by a long flexible body, shearing teeth and forward facing eyes. The largest family of Carnivora is the Mustelidae, including 67 species of weasels, martens, minks, skunks, badgers, and otters, although the Canidae (dogs and wolves) and Felidae (cats, tigers and lions) are more prominent. Canine teeth and powerful jaws assist in the ability to hunt successfully, and most carnivores are either ground - dwelling or skilled climbers (Nowak 1991). Annual litters are vary widely in size (Janson and Pope, 1989).

Most carnivores are solitary although some Canidae and Hyaenidae hunt in packs and communicate through barks and growls. Bears (Ursidae), Raccoon-like animals (Procyonidae), weasels and related animals (Mustelidae), and civets and mongooses (Viverridae) are solitary. A few species of Felidae live in prides of several related females and their young along with one or more unrelated males (Janson and Pope, 1989). Extended parental care of young is delivered in most carnivores.

PINNIPEDIA (seals and sea lions)

Pinnipeds include the ear-less seals (Phocidae, 18 species), the eared-seals, sea lions and fur seals (Otariidae, 16 species), and the walrus (Odobenidae, 1 species). One of the two orders of marine mammals, they are a monophyletic group most closely related to the Carnivora (Nowak 1991). They have streamlined bodies with both hair and sub-dermal blubber . The forelimbs provide the power for underwater movement in the Otariidae and Odobenidae who are also able to move quickly across land. The Phocidae use the hindlimb flippers for forward impulsion and are clumsy on land. Sexual dimorphism is common in the Otariids and Odobenids with some males (e.g. southern elephant seals) weighing as much as four times as much as females (Janson and Pope, 1989). Walruses are native to the Arctic Ocean, from which they migrate in the winter, and are characterized by a thick durable skin and tusks in both males and females. Male walrus may measure up to 13 feet and may weigh as much as 3000 pounds (Janson and Pope, 1989).

Pinnipeds eat fish, crustaceans, molluscs, and other invertebrates. They exhibit highly seasonal mating, delayed implantation and gestation periods of 8-11 months (Eisenberg 1981). They bear a single pup that can swim almost from birth (Janson and Pope, 1989).

TUBULIDENTATA (aardvarks)

Aardvarks feed nocturnally on termites and ants and sleep inside burrows by day. The aardvark is found on African savannas where they sometimes travel long distances in search of food (Janson and Pope, 1989). Most aardvarks bear single offspring (Nowak 1991).

PROBOSCIDEA (elephants)

There are only two species in this order, the African and the Indian Elephants, remnants of a much more diverse order in the Pliocene. African elephants are the larger and both males and females bear tusks (continuously growing upper incisors). Indian elephants have one more toe on the hindfoot and only the males have tusks. Both feed for much of the day on grasses, bark, roots, and leaves to support their massive bodies (Janson and Pope, 1989). Using the trunk to lift food and water, elephants rely on grinding teeth to masticate woody fibers.

Both species have a social organization based on matriarchal groups of females and their young with males associated for shorter lengths of time. Gestation in the Indian elephant is approximately 646 days in the Indian elephant; young nurse for 18 months and females first give birth at 15-16 years (Nowak 1991). Female African elephants appear to select older "musth" males for mating, have a 22 month gestation, and young usually nurse 6-18 months with observations as long as 6 years. Young are produced about every 5 years.

HYRACOIDEA (hyraxes)

The highly sociable Hyraxes look like large guinea pigs, but are most closely related to elephants. As in elephants, grinding teeth serve to assist in digestion of plants, and long incisors are externally visible. Rock-dwelling species tend to be more social than arboreal species (Janson and Pope, 1989). Homeothermic control is poor (Nowak 1991). Social groups include a territorial male and several females. Gestation is somewhat over 200 days and young nurse for 1 to 5 months. All hyraxes engage in an elaborate system of communication consisting of noises and cries including predator alerts.

SIRENIA (manatees, dugongs and sea cows)

Sirenia are tropical and subtropical herbivorous aquatic animals of coastal waters and rivers. Sirenia predate the related Proboscidea, with their maximal abundance in the Oligocene to Pliocene. They are characterized by a large body with a thick layer of blubber, may weigh up to 650 kg and grow to lengths of 4 meters (Janson and Pope, 1989). Young use forelimbs for propulsion but adults use the fluke-like tail. All Sirenia have cheek teeth for grinding and some species (eg. Dugong) also have incisors. Sirenians are generally solitary or found in small groups. Females give birth to a solitary young after a gestation of 13 to 14 months. Lactation lasts for 18 months and sexual maturity occurs between 10 and 15 years (Nowak 1991).

PERISSODACTYLA (odd toed ungulates)

Ungulates are horned and hoofed herbivores which are characterized by their ability to chew and digest fibrous plants, and to run swiftly. Most ungulates breed annually and bear single offspring (Janson and Pope, 1989). Gestation requires roughly a year. The Perissodactyla are single or three-toed ungulates native to Africa, parts of Asia and Central and South America. Seventeen recent species are included in three families of horses (Equidae), tapirs (Tapiridae) and rhinoceroses (Rhinocerotidae), while an additional family became extinct in the Pleistocene (Nowak 1991). They are grazers of fibrous plant material digested by grinding teeth and simple stomachs. Some Equiidae live in migratory herds on the plains in single male, multi-female groups. These horses, zebras and asses have interbirth intervals of 1 to 3 years. Tapiridae are shorter-legged herbivores native to Malaysia, Asia, Sumatra, Java, and Central and South America where they live in dense underbrush. Rhinocerotidae are native to Africa, Asia, Sumatra, and Borneo; they feed on vegetation in savanna and wooded areas (Janson and Pope, 1989). Gestation exceeds one year and adults have few predators beyond humans.

ARTIODACTYLA (even toed ungulates).

Nine families of Artiodactyls include the hog-like animals (Sulidae), peccaries (Tayassuidae), hippopotamuses (Hippopotamidae), camels and llamas (Camelidae), chevrotains (Tragulidae), giraffes (Giraffidae), deer-like animals (Cervidae), cattle, antelope, goats, and sheep (Bovidae), and pronghorn (Antilocapridae). The Artiodactyla have four toes.

Artiodactyla are predominantly herbivores (though the Suidae are omnivores) with a two-chambered stomach. Suidae tend to live in forest habitats, feeding on roots, plants, bird eggs, and small animals nocturnally. Artiodactyls tend to travel in groups or herds of up to 50 individuals. Artiodactyl species inhabit savannah, forest, riverene and desert habitats.

References Cited

Nowak RM. *Walker's Mammals of the World Fifth Edition*, The Johns Hopkins University Press, Baltimore, 1991.

Grant, T. R. and F. N. Carrick, 1978. "Some aspects of the ecology of the platypus, Ornithorhyncus anatinus, in the upper Shoalhaven River, New South Wales." *Australian Journal of Zoology* 20: 181-199.

Janson, M. and J. Pope, Eds., 1989. *The Animal World.* Chicago, World Book, Inc.

McKenna, M. C., 1975. Toward a phylogenetic classification for the Mammalia. in: *Phylogeny of the Primates: A Multidisciplinary Approach.* W. P. Luckett and F. S. Szalay. New York, Plenum Press: 21-46.

Nowak, R. M., 1991. *Walker's Mammals of the World, Vol. I.* Baltimore, Johns Hopkins University Press.

Whitehead, H., 1995. "The realm of the elusive sperm whale." *National Geographic* (November): 58-73.

Table 1. Record Life Spans (years) of Mammals

((+) indicates that individual was still alive at the time that life span was recorded)

Order/Family	Genus/Species	Common Name	Wild	Capt.	M/F	Reference
Artiodactyla						
Antilocapridae	Antilocapra americana	Pronghorn	10.0		x	[100]
	Antilocapra americana	Pronghorn		11.8	x	[143]
	Antilocapra americana	Pronghorn	10.0	12.0	x	[192]
Bovidae	Addax nasomaculatus	Addax		19.0	x	[100]
	Addax nasomaculatus	Addax		25.3	x	[143]
	Aepyceros melampus	Impala	15.0		x	[100]
	Aepyceros melampus	Impala (+)	13.0	17.4	x	[134] [155] [143]
	Alcelaphus buselaphus	Hartebeest		19.0	x	[100]
	Alcelaphus buselaphus	Hartebeest	20.0		x	[181] [292]
	Ammotragus lervia	Barbary sheep	10.0	20.0	x	[100]
	Ammotragus lervia	Barbary sheep		20.9	x	[143]
	Antidorcas marsupialis	Springbok	20.0		x	[100]
	Antidorcas marsupialis	Springbok		19.0	x	[143]
	Antilope cervicapra	Blackbuck	15.0		x	[100]
	Antilope cervicapra	Blackbuck (+)	18.0		f	[251]
	Bison bison	American bison	33.0		x	[21]
	Bison bison	American bison	25.0		x	[100]
	Bison bonasus	European bison	27.0		x	[100]
	Bison spp.	Bison	20.0	40.0	x	[10] [101] [112] [167] [188] [200]
	Bos gaurus	Seladang		26.2	x	[143]
	Bos grunniens	Yak	25.0		f	[101]
	Bos taurus	Domestic cattle (+)		20.0	x	[229]
	Boselaphus tragocamelus	Nilgai		21.0	x	[100]
	Boselaphus tragocamelus	Nilgai		21.7	x	[143]
	Bubalus arnee	Wild water buffalo	25.0		x	[100]
	Bubalus depressicornis	Lowland anoa	30.0		x	[100]
	Bubalus mindorensis	Asian water buffalo		28.0	x	[101] [243]
	Budorcas taxicolor	Takin		15.8	x	[143]
	Capra aegagrus	Wild goat	12.0		x	[100]
	Capra cylindricornis	East Caucasian tur	22.0		x	[229]
	Capra falconeri	Markhor	13.0		x	[100]
	Capra falconeri	Markhor	12.0		x	[229]
	Capra hircus domesticus	Domestic goat	20.8		x	[21]
	Capra ibex	Ibex		22.3	x	[143]
	Capra ibex cylindricornis	East Caucasian tur	12.0		x	[100]
	Capra ibex ibex	Alpine ibex	14.0		x	[100]
	Capra ibex nubiana	Nubian ibex		17.0	x	[100]
	Capra ibex sibirica	Asiatic ibex	16.0		x	[100]
	Capra pyrenaica	Spanish ibex	16.0		x	[100]

Capricornis crispus	Serow (+)		10.0	x	[176] [306] [310] [335]
Capricornis sumatraensis	Serow		21.0	x	[100]
Cephalophus monticola	Blue duiker	12.0		x	[100]
Cephalophus nigrifrons	Duiker		19.7	x	[143]
Connochaetes gnou	Black wildebeest		20.0	x	[100]
Connochaetes taurinus	Brindled wildebeest		20.0	x	[100]
Connochaetes taurinus	Wildebeest		21.5	x	[143]
Damaliscus dorcas	Sassaby	21.4		x	[143]
Damaliscus dorcas dorcas	Bontebok		17.0	x	[100]
Damaliscus dorcas phillipsi	Blesbok		17.0	x	[100]
Gazella dorcas	Dorcas gazelle	12.5		x	[100]
Gazella dorcas	Dorcas gazelle		17.1	x	[143]
Gazella gazella	Mountain gazelle	12.0		x	[100]
Gazella leptoceros	Slender-horned gazelle	14.0		x	[100]
Gazella soemmeringi	Soemmering's gazelle	14.0		x	[100]
Gazella subgutturosa	Persian gazelle	12.0		x	[100]
Gazella thomsoni	Thomson's gazelle	10.5		x	[100]
Hemitragus hylocrius	Nilgiri tahr	9.0		x	[100]
Hemitragus jayakari	Arabian tahr	14.0		x	[100]
Hemitragus jemlahicus	Himalayan tahr	10.0		x	[100]
Hemitragus jemlahicus	Himalayan tahr		21.8	x	[143]
Hippotragus equinus	Roan antelope		17.0	x	[100]
Hippotragus niger	Sable antelope		17.0	x	[100]
Hippotragus niger	Sable antelope		19.8	x	[143]
Kobus ellipsiprymnus	Waterbuck		18.0	x	[100]
Kobus ellipsiprymnus	Waterbuck		18.7	x	[143]
Kobus ellipsiprymnus	Waterbuck	18.5		f	[299]
Kobus kob	Kob		17.0	x	[100]
Kobus leche	Lechwe waterbuck	15.0		x	[100]
Kobus megaceros	Mrs. Gray's waterbuck	10.0		x	[100]
Litocranius walleri	Gerenuk	12.0		x	[100]
Litocranius walleri	Gerenuk (+)	8.0	13.0	x	[178, 179]
Madoqua (= Rhynchotragus) kirki	Kirk's dik-dik	10.0		x	[52] [114] [152]
Madoqua (= Rhynchotragus) kirki	Kirk's dik-dik		10.0	x	[100]
Nemorhaedus goral	Goral	15.0		x	[100]
Nemorhaedus spp.	Goral		17.6	x	[143]
Neotragus moschatus	Suni		9.0	x	[100]
Neotragus moschatus	Suni		10.2	x	[143]
Neotragus pygmaeus	Royal antelope		6.0	x	[100]
Oreamnos americanus	Mountain goat	18.0		x	[100]
Oreamnos americanus	Mountain goat	14.0		m	[229]
Oreamnos americanus	Mountain goat	18.0		f	[229]
Oreotragus oreotragus	Klipspringer		15.0	x	[100]
Oreotragus oreotragus	Klipspringer		12.1	x	[143]
Oryx gazella	S. African oryx antelope		20.0	x	[100]
Oryx spp.	Oryx	20.0		x	[229]
Ourebia ourebi	Oribi		14.0	x	[100] [161]
Ovibos moschatus	Muskox	24.0		x	[4] [100]

	Ovis ammon	Argali	13.0		x	[100]
	Ovis canadensis	American bighorn sheep	24.0		x	[192]
	Ovis orientalis	Asiatic mouflon	13.0		x	[100]
	Ovis orientalis	Urial	6.0		x	[192]
	Ovis spp.	Sheep	20.0		m	[88]
	Ovis spp.	Sheep	24.0		f	[88]
	Pelea capreolus	Rhebok	10.0		x	[101]
	Procapra gutturosa	Central Asian gazelle		7.0	x	[143]
	Pseudois nayaur	Bharal		15.0	x	[100]
	Pseudois nayaur	Bharal	24.0		x	[192]
	Pseudois spp.	Bharal		20.3	x	[143]
	Redunca arundinum	Southern reedbuck		10.0	x	[100]
	Redunca fulvorufula	Mountain reedbuck	12.0		x	[100]
	Redunca redunca	Bohor reedbuck		10.0	x	[100]
	Redunca redunca	Bohor reedbuck		18.0	x	[143]
	Rupicapra rupicapra	Chamois	20.0		x	[100]
	Rupricapra spp.	Chamois	22.0		x	[229]
	Saiga tatarica	Saiga antelope	10.0		x	[100]
	Saiga tatarica	Saiga antelope	12.0		x	[229]
	Sylvicapra grimmia	Gray duiker		12.0	x	[100]
	Sylvicapra grimmia	Gray duiker		14.3	x	[143]
	Syncerus caffer	African buffalo	29.5		x	[21]
	Syncerus caffer	African buffalo	26.0	16.0	x	[100]
	Syncerus caffer	African buffalo	18.0	29.5	x	[143]
	Taurotragus oryx	Eland		23.6	f	[311]
	Tetracerus quadricornis	Four-horned antelope	10.0		x	[100]
	Tetracerus quadricornis	Four-horned antelope		10.0	x	[143]
	Tragelaphus angasi	Nyala	16.0		x	[100]
	Tragelaphus euryceros	Bongo	19.0		x	[100]
	Tragelaphus euryceros	Bongo		19.4	f	[252]
	Tragelaphus imberbis	Lesser kudu (+)	15.0		x	[100]
	Tragelaphus imberbis	Lesser kudu (+)	10.0	15.0	x	[177] [180]
	Tragelaphus oryx	Eland antelope	25.0		x	[100]
	Tragelaphus spekei	Sitatunga	20.0		x	[161]
	Tragelaphus spekei	Sitatunga	19.0		x	[100]
	Tragelaphus strepsiceros	Greater kudu		23.0	x	[100]
	Tragelaphus strepsiceros	Greater kudu		20.8	x	[143]
Camelidae	Camelus dromedarius	Dromedary, one-humped camel	40.0		x	[100]
	Camelus ferus	Bactrian camel (+)	35.0		x	[21]
	Camelus ferus	Bactrian, two-humped camel	40.0		x	[100]
	Camelus spp.	Camel		50.0	x	[87]
	Lama guanicoe	Guanaco	20.0		x	[100]
	Lama guanicoe	Guanaco		28.3	x	[143]
	Lama guanicoe f. glama	Llama	20.0		x	[100]
	Vicugna vicugna	Vicuna	20.0		x	[100]
	Vicugna vicugna	Vicuna		24.8	x	[78, 79] [277]
Cervidae	Alces alces	Moose (+)	16.0		x	[100]
	Alces alces	Moose	27.0		x	[241]
	Axis axis	Axis deer		15.0	x	[100]
	Axis axis	Axis deer		20.8	x	[143]

	Axis porcinus	Hog deer		20.0	x	[100]
	Capreolus capreolus	Roe deer	15.0		x	[100]
	Capreolus spp.	Roe deer	17.0		x	[229]
	Cervus elaphus	Red deer	26.8		x	[21]
	Cervus elaphus	Red deer	20.0		x	[100]
	Cervus elaphus	Red deer		24.7	x	[143]
	Cervus eldi	Brow-antlered deer		19.4	f	[246]
	Cervus (Przewalskium) albirostris	Thorold's deer	18.0		x	[100]
	Dama dama	Fallow deer (+)		20.0	x	[72]
	Dama dama	Fallow deer	25.0		x	[100]
	Elaphurus davidianus	Pere David's deer	18.0		x	[100]
	Elaphurus davidianus	Pere David's deer	23.3		x	[144]
	Hippocamelus antisensis	Guemals		10.6	x	[143]
	Hydropotes inermis	Chinese water deer	12.0		x	[100]
	Hydropotes inermis	Chinese water deer (+)		11.5	x	[143]
	Mazama americana	Red brocket	12.0		x	[100]
	Mazama americana	Red brocket		13.8	x	[143]
	Mazama gouazoubira	Brown brocket	12.0		x	[100]
	Moschus spp.	Musk deer		20.0	x	[95]
	Muntiacus muntjak	Indian muntjac (+)		17.0	x	[100]
	Muntiacus muntjak	Indian muntjac (+)		17.6	x	[143]
	Odocoileus virginianus	White-tailed deer	10.0	16.0	x	[100]
	Odocoileus virginianus	White-tailed deer		23.0	x	[312]
	Pudu mephistopheles	Northern pudu	10.0		x	[100]
	Pudu pudu	Southern pudu		12.5	x	[143]
	Pudu pudu	Southern pudu	10.0		x	[100]
	Rangifer tarandus	Caribou	15.0	20.2	x	[10] [16] [273] [143]
	Rangifer tarandus	Caribou (+)	8.0		m	[100]
	Rangifer tarandus	Caribou (+)	10.0		f	[100]
Giraffidae	Giraffa camelopardalis	Giraffe	36.3		x	[21]
	Giraffa camelopardalis	Giraffe	26.0		f	[49]
	Giraffa camelopardalis	Giraffe	25.0	28.0	x	[100] [192]
	Giraffa camelopardalis	Giraffe		36.2	x	[143]
	Okapia johnstoni	Okapi (+)		30.0	x	[100]
	Okapia johnstoni	Okapi		33.0	x	[143]
	Okapia johnstoni	Okapi		15.0	x	[192]
Hippopotamidae	Choeropsis liberiensis	Pygmy hippopotamus	42.0		x	[100]
	Choeropsis liberiensis	Pygmy hippopotamus		43.8	x	[143]
	Choeropsis liberiensis	Pygmy hippopotamus	35.0	42.0	x	[192]
	Hippopotamus amphibius	Hippopotamus	54.5		x	[21]
	Hippopotamus amphibius	Hippopotamus	50.0	40.0	x	[100]
	Hippopotamus amphibius	Hippopotamus		54.3	x	[143]
	Hippopotamus amphibius	Hippopotamus	45.0	49.0	x	[192]
Suidae	Babyrousa babyrussa	Babirusa		24.0	x	[24] [54] [101]
	Babyrousa babyrussa	Babirusa	24.0		x	[100] [192]
	Hylochoerus meinertzhageni	Giant forest hog		3.0	x	[100]
	Phacochoerus aethiopicus	Wart hog	18.8		x	[100]
	Phacochoerus aethiopicus	Wart hog		18.8	x	[143]

	Phacochoerus aethiopicus	Wart hog	15.0		x	[192]
	Potamochoerus porcus	Bush pig	20.0		x	[100]
	Potamochoerus porcus	Bush pig		20.0	x	[143]
	Potamochoerus porcus	Bush pig	15.0		x	[192]
	Sus salvanius	Pygmy hog	7.5		x	[100]
	Sus salvanius	Pygmy hog	12.0		x	[192]
	Sus scrofa	Wild boar	21.0		x	[100]
	Sus scrofa	Wild boar	20.0		x	[192]
	Sus verrucosus	Javan pig	14.0		x	[100]
	Sus verrucosus (= celebensis)	Javan pig (+)	10.0		x	[100]
	Sus spp.	Pig, hog, or boar		27.0	x	[229]
Tayassuidae	Catagonus wagneri	Chaco peccary		1.0	x	[100]
	Tayassu pecari	White-lipped peccary	13.3		x	[100]
	Tayassu tajacu	Collared peccary	24.7		x	[100]
	Tayassu tajacu	Collared peccary		24.6	x	[143]
	Tayassu tajacu	Collared peccary	10.0	21.0	x	[192]
Tragulidae	Hyemoschus aquaticus	Water chevrotain	13.0		x	[100]
	Hyemoschus aquaticus	Water chevrotain	14.0		x	[192]
	Tragulus javanicus	Lesser Malay chevrotain		12.0	x	[100]
	Tragulus napu	Larger Malay chevrotain		14.0	x	[100]
	Tragulus napu	Larger Malay chevrotain (+)		14.0	x	[143]
Carnivora						
Ailuridae	Ailurus fulgens	Lesser panda	14.0		x	[100]
	Ailurus fulgens	Lesser panda		13.4	x	[143]
Ailuropodidae	Ailuropoda melanoleuca	Giant panda	26.0		x	[21]
	Ailuropoda melanoleuca	Giant panda		30.0	x	[100]
Canidae	Alopex lagopus	Arctic fox	10.0		x	[100]
	Alopex lagopus	Arctic fox		15.0	x	[143]
	Atelocynus microtis	Small-eared dog		11.0	x	[143]
	Canis adustus	Side-striped jackal		10.0	x	[266]
	Canis aureus	Golden jackal		14.0	x	[100]
	Canis aureus	Golden jackal		16.0	x	[159] [176] [216, 217] [228] [268] [321]
	Canis dingo	Dingo		14.0	x	[192]
	Canis familiaris dingo	Dingo	14.8		x	[101]
	Canis familiaris	Domestic dog	29.5		x	[21]
	Canis familiaris	Domestic dog		20.0	x	[8] [32] [100]
	Canis latrans	Coyote	14.5	21.8	x	[13] [90] [153] [143]
	Canis latrans	Coyote	14.5	18.0	x	[192]
	Canis latrans	Coyote (+)	21.0		x	[21]
	Canis latrans	Coyote (+)		15.0	x	[100]
	Canis lupus	Gray wolf	10.0	20.0	x	[100]
	Canis lupus	Gray wolf	16.0	20.0	x	[192]
	Canis lupus	Gray wolf	13.7	16.0	f	[201-206] [317, 318]
	Canis lupus f. dingo	Dingo (+)		14.0	x	[100]

	Canis mesomelas	Black-backed jackal		14.0	x	[12] [22] [159] [215, 216]
	Canis rufus	Red wolf		14.0	x	[37]
	Canis rufus	Red wolf	16.0	20.0	x	[192]
	Cerdocyon thous	Crab-eating fox		11.5	x	[294]
	Chrysocyon brachyurus	Maned wolf		15.0	x	[100] [192]
	Chrysocyon brachyurus	Maned wolf		13.0	x	[143]
	Cuon alpinus	Dhole		16.0	x	[229]
	Fennecus zerda	Fennec fox		12.0	x	[100]
	Fennecus zerda	Fennec fox		11.6	x	[143]
	Lycaon pictus	African hunting dog	11.0		x	[77] [159] [193] [275] [319] [321]
	Lycaon pictus	African hunting dog	10.0		x	[100] [192]
	Nyctereutes procyonoides	Raccoon dog	11.0	7.0	x	[100]
	Nyctereutes procyonoides	Raccoon dog		10.7	x	[143]
	Otocyon megalotis	Bat-eared fox		13.8	x	[143]
	Otocyon megalotis	Bat-eared fox (+)		6.0	x	[100]
	Pseudalopex gymnocercus	S. American fox		13.7	x	[143]
	Speothos venaticus	Bush dog		10.3	x	[143]
	Speothos venaticus	Bush dog (+)		10.0	x	[100]
	Urocyon cinereoargenteus	Gray fox (+)		13.0	x	[100]
	Urocyon spp.	Gray fox		13.7	x	[143]
	Vulpes bengalensis	Indian fox (+)	10.0		x	[100]
	Vulpes cana	Blanford's fox	10.0		x	[100]
	Vulpes chama	Cape fox	10.0		x	[100]
	Vulpes macrotis	Kit fox (+)		20.0	x	[143]
	Vulpes pallida	Pale fox	10.0		x	[100]
	Vulpes velox	Swift fox		12.8	x	[59] [156]
	Vulpes velox	Swift fox		12.0	x	[100]
	Vulpes vulpes	Red fox	7.0	15.0	x	[1]
	Vulpes vulpes	Red fox	12.0		x	[1]
Felidae	Acinonyx jubatus	Cheetah		19.0	x	[57] [101] [159]
	Acinonyx jubatus	Cheetah	15.0		x	[100]
	Acinonyx jubatus	Cheetah	12.0	17.0	x	[192]
	Felis caracal	Caracal		17.0	x	[159]
	Felis catus	Domestic cat	34.0		x	[21]
	Felis (= Profelis = Puma) concolor	Cougar (+)		20.0	x	[47]
	Felis (= Lynx) lynx	Lynx		26.8	x	[257]
	Felis nigripes	Black-footed cat	12.0		x	[100]
	Felis (= Leopardus) pardalis	Ocelot		20.3	x	[143]
	Felis (= Lynx) rufus	Bobcat	32.3		x	[21]
	Felis (= Lynx) rufus	Bobcat		32.3	x	[143]
	Felis (= Leptailurus) serval	Serval		19.8	f	[323]
	Felis silvestris	Wild cat	18.0		x	[100]
	Felis silvestris	Wild cat	15.0		x	[192]
	Felis silvestris	Wild cat		15.0	x	[159]
	Felis (= Profelis) temmincki	Asian golden cat		18.0	x	[176]

	Felis (= Leopardus) wiedii	Margay (+)		13.0	x	[101]
	Leopardus tigrinus	Little spotted cat		20.0	x	[100]
	Leopardus wiedii	Margay (+)		20.0	x	[100]
	Leptailurus serval	Serval	23.0		x	[100]
	Lynx (= Felis) rufus	Bobcat	15.0		x	[100]
	Neofelis nebulosa	Clouded leopard		17.0	x	[192] [100, 101]
	Neofelis (= Panthera) tigris	Tiger	18.0	25.0	x	[100]
	Panthera leo	African lion		30.0	x	[101, 102] [159] [275]
	Panthera leo	Lion	29.0		x	[21]
	Panthera leo	Lion	15.0	25.0	x	[100]
	Panthera leo	Lion	15.0	24.0	x	[192]
	Panthera onca	Jaguar		22.0	x	[101, 102] [218] [100]
	Panthera onca	Jaguar		20.0	x	[192]
	Panthera pardus	Leopard	12.0	20.0	x	[192]
	Panthera pardus	Leopard (+)		23.0	x	[101, 102] [159]
	Panthera tigris	Tiger	26.3		x	[21]
	Panthera tigris	Tiger	15.0	20.0	x	[192]
	Panthera tigris	Tiger	26.0	26.0	x	[197]
	Panthera uncia	Snow leopard		15.0	x	[192]
	Panthera uncia	Snow leopard		19.0	x	[133]
	Prionailurus bengalensis	Leopard cat	15.0		x	[100]
	Profelis (= Caracal) caracal	Caracal (+)	12.0		x	[100]
	Profelis (= Puma) concolor	Mountain lion	18.0		x	[100]
	Uncia (= Panthera) uncia	Snow leopard		18.0	x	[100]
Hyaenidae	Crocuta crocuta	Spotted hyena		41.1	x	[143]
	Crocuta crocuta	Spotted hyena	25.0	40.0	x	[192]
	Hyaena brunnea	Brown hyena	12.0		x	[211]
	Hyaena brunnea	Brown hyena		13.0	x	[192]
	Hyaena hyaena	Striped hyena		24.0	x	[259, 260]
	Hyaena hyaena	Striped hyena		24.0	x	[192]
	Proteles cristatus	Aardwolf		14.3	x	[143]
	Proteles cristatus	Aardwolf		13.0	x	[192]
Mustelidae	Aonyx capensis	Cape clawless otter		11.0	x	[100] [143]
	Arctonyx collaris	Hog badger		13.9	x	[143] [100]
	Conepatus chinga	Hog-nosed skunk		6.6	x	[143]
	Conepatus leuconotus	Hog-nosed skunk		7.0	x	[100]
	Eira barbara	Tayra		18.0	x	[100, 101]
	Enhydra lutris	Sea otter	23.0		f	[271]
	Enhydra lutris	Sea otter		19.0	m	[271]
	Enhydra lutris	Sea otter	30.0		x	[100]
	Galictis vittata	Grison		10.5	x	[143]
	Gulo gulo	Wolverine		17.3	x	[143]
	Gulo gulo	Wolverine		18.0	x	[100]
	Ictonyx striatus	Striped polecat, zorilla		13.3	x	[143] [100]
	Lutra canadensis	Canadian otter	21.0		x	[21]
	Lutra canadensis	River otter		23.0	f	[229]
	Lutra lutra	Eurasian river otter		22.0	x	[100]

	Martes americana	American marten		17.0	x	[100] [229]
	Martes flavigula	Yellow-throated marten		14.0	x	[176] [304] [100]
	Martes foina	Beech marten		18.1	x	[143]
	Martes martes	European pine marten		17.0	x	[101] [304]
	Martes martes	Pine marten	10.0	15.0	x	[100]
	Martes pennnti	Fisher	10.0	10.0	x	[244 , 245] [100]
	Martes zibellina	Sable		15.0	x	[101] [228] [304] [100]
	Meles meles	Badger		16.0	x	[100]
	Meles meles	Old World badger		16.2	x	[143]
	Mellivora capensis	Honey badger	26.4		x	[143]
	Mellivora capensis	Honey badger		26.5	x	[100]
	Melogale moschata	Ferret badger (+)		10.5	x	[143]
	Mephitis mephitis	Striped skunk	6.0	10.0	x	[100]
	Mephitis spp.	Striped skunk		12.9	x	[143]
	Mustela lutreola	European mink	10.0		x	[228] [304] [100]
	Mustela nigripes	Black-footed ferret		12.0	x	[117] [100]
	Mustela putorius	European polecat	6.0	14.0	x	[20] [100] [228] [304]
	Mustela putorius	European polecat (+)	10.0		x	[100]
	Mustela sibirica	Siberian weasel		8.8	x	[143] [100]
	Mustela vison	American mink	10.0		x	[10] [35] [132] [191] [100]
	Poecilictis libyca	N. African striped weasel (+)		5.0	x	[268] [100]
	Poecilogale albinucha	Striped weasel		5.2	x	[296]
	Poecilogale albinucha	Striped weasel		5.0	x	[100]
	Pteronura brasiliensis	Giant otter		12.8	x	[143]
	Pteronura brasiliensis	Giant otter		12.0	x	[100]
	Spilogale putorius	Spotted skunk		10.0	x	[100]
	Spilogale spp.	Spotted skunk		9.8	x	[60]
	Taxidea taxus	American badger	14.0	26.0	x	[143]
	Taxidea taxus	American badger	26.0		x	[21]
	Taxidea taxus	American badger	14.0	16.0	x	[100]
	Vormela peregusna	Marbled polecat		8.9	x	[143] [100]
Procyonidae	Bassaricyon gabbii	Olingo (+)	24.0		x	[100]
	Bassaricyon spp.	Olingo		17.1	x	[143]
	Bassariscus astutus	Ringtail	16.0		x	[100]
	Bassariscus astutus	Ringtail		14.3	x	[143]
	Bassariscus sumichrasti	Cacomistle	23.0		x	[100]
	Nasua nasua (= narica)	Coatimundi	14.0		x	[100]
	Nasua nasua	Coatimundi		17.7	x	[143]
	Nasua nasua	Coatimundi	14.0		x	[100]
	Potos flavus	Kinkajou	29.0		x	[100]
	Potos flavus	Kinkajou		23.6	x	[129] [229]
	Procyon cancrivorus	Crab-eating raccoon	14.0		x	[100]
	Procyon lotor	Common raccoon (+)	20.0		x	[100]
	Procyon spp.	Common raccoon	16.0	20.6	x	[143]

Ursidae	Melursus (= Ursus) ursinus	Sloth bear		30.0	x	[100]
	Tremarctos ornatus	Spectacled bear		25.0	x	[100]
	Tremarctos ornatus	Spectacled bear		36.4	x	[143]
	Ursus (= Euarchtos) americanus	American black bear	26.0		f	[10] [145] [242]
	Ursus (= Euarctos) americanus	American black bear	32.0		x	[192]
	Ursus arctos	Grizzly bear	25.0	50.0	x	[46] [58] [91] [145] [221] [237] [293] [304]
	Ursus arctos	Grizzly bear	30.0	47.0	x	[192]
	Ursus malayanus	Malayan sun bear		24.8	x	[143]
	Ursus (= Thalarctos) maritimus	Polar bear (+)	30.0	38.2	x	[51] [253] [300] [316] [169]
	Ursus (= Thalarctos) maritimus	Polar bear	25.0		x	[192]
	Ursus (= Selenarctos) thibetanus	Asiatic black bear		33.0	x	[176]
	Ursus (= Melursus) ursinus	Sloth bear		40.0	x	[307] [171] [229]
	Ursus ursus arctos	European brown bear	47.0		x	[21]
Viverridae	Arctictis binturong	Binturong	18.0		x	[100]
	Arctictis binturong	Binturong		22.7	x	[143]
	Arctogalidia trivirgata	Small-toothed palm civet		15.8	x	[143]
	Arctogalidia trivirgata	Small-toothed palm civet	12.0		x	[100]
	Atilax paludinosus	Marsh mongoose		17.4	x	[143]
	Atilax paludinosus	Marsh mongoose (+)		17.0	x	[100]
	Bdeogale nigripes	Black-legged mongoose (+)		15.0	x	[100]
	Bdeogale nigripes	Black-legged mongoose		15.8	x	[143]
	Civettictis civetta	African civet		28.0	x	[143]
	Civettictis civetta	African civet (+)	15.0		x	[100]
	Crossarchus obscurus	Dark mongoose		4.5	x	[100]
	Crossarchus spp.	Cusimanse		9.0	x	[93]
	Cryptoprocta ferox	Fossa (+)		20.0	x	[2] [3] [229] [164]
	Cynogale bennettii	Otter civet (+)		5.0	x	[100]
	Cynogale bennettii	Otter civet		5.0	x	[143]
	Cynictis penicillata	Yellow mongoose (+)		15.0	x	[100]
	Cynictis penicillata	Yellow mongoose		15.2	x	[143]
	Fossa fossa	Malagasy civet		11.0	x	[143]
	Fossa fossa	Malagasy civet		11.0	x	[100]
	Galidia elegans	Malagasy ring-tailed mongoose (+)		13.0	x	[100]
	Galidia elegans	Malagasy ring-tailed mongoose		13.2	x	[143]
	Genetta genetta	European genet		34.0	x	[192]
	Genetta genetta	European genet (+)		14.0	x	[100]
	Genetta genetta	European genet		21.6	x	[143]
	Genetta genetta	European genet		13.0	f	[159] [268] [330]
	Genetta tigrina	Large-spotted genet		9.5	x	[100]
	Helogale parvula	Dwarf mongoose		10.0	x	[100]
	Hemigalus derbyanus	Banded palm civet (+)		12.0	x	[143] [100]

	Herpestes ichneumon	Mongoose	12.0		x	[100]
	Herpestes ichneumon	Mongoose (+)		20.0	x	[67] [100] [159] [176] [263]
	Herpestes javanicus	Javan mongoose (+)		8.0	x	[100]
	Herpestes sanguineus	Slender mongoose		6.0	x	[100]
	Ichneumia albicauda	White-tailed mongoose (+)		12.0	x	[100]
	Ichneumia albicauda	White-tailed mongoose (+)		10.0	x	[143]
	Mungos mungo	Banded mongoose		12.0	x	[322]
	Mungos mungo	Banded mongoose		11.0	x	[100]
	Nandinia binotata	African palm civet		15.8	x	[143]
	Nandinia binotata	African palm civet (+)		15.0	x	[100]
	Paguma larvata	Masked palm civet		15.4	x	[176] [207]
	Paguma larvata	Masked palm civet		15.0	x	[192]
	Paguma larvata	Masked palm civet	18.0		x	[100]
	Paradoxurus hermaphroditus	Common palm civet (+)	22.0		x	[100]
	Paradoxurus hermaphroditus	Common palm civet		22.4	x	[207]
	Poiana richardsoni	African linsang		5.3	x	[143]
	Poiana richardsoni	African linsang (+)		5.0	x	[100]
	Prionodon linsang	Banded linsang (+)		10.0	x	[100]
	Prionodon linsang	Banded linsang		10.7	x	[143]
	Salanoia concolor	Salano	4.8		x	[21]
	Salanoia concolor	Salano		4.8	x	[143]
	Suricata suricatta	Meerkat		12.5	x	[143]
	Suricata suricatta	Meerkat (+)		12.0	x	[100]
	Viverra civetta	African civet	28.0		x	[21]
	Viverra tangalunga	Tangalunga		12.0	x	[100]
	Viverra zibetha	Oriental civet (+)		15.0	x	[100]
	Viverra zibetha	Oriental civet (+)		20.0	x	[294]
	Viverricula indica	Lesser Oriental civet		10.5	x	[143]
	Viverricula indica	Lesser Oriental civet (+)		8.0	x	[100]
Cetacea						
Balaenidae	Balaena mysticetus	Bowhead whale	40.0		x	[10] [26] [65] [195] [226]
	Balaena mysticetus	Bowhead whale (+)	30.0		x	[100]
Balaenopteridae	Balaenoptera acutorostrata	Minke whale	50.0		x	[100]
	Balaenoptera acutorostrata	Minke whale	47.0		x	[105]
	Balaenoptera acutorostrata	Minke whale	45.0		x	[192]
	Balaenoptera borealis	Sei whale	70.0		x	[100]
	Balaenoptera borealis	Sei whale	74.0		x	[105]
	Balaenoptera edeni	Bryde's whale	72.0		x	[65] [105] [185] [213] [284]
	Balaenoptera musculus	Blue whale	45.0		x	[21]
	Balaenoptera musculus	Blue whale	50.0		x	[100]
	Balaenoptera musculus	Blue whale	110.		x	[105]
	Balaenoptera musculus	Blue whale	65.0		x	[192]
	Balaenoptera physalus	Fin whale	116.		x	[21]
	Balaenoptera physalus	Fin whale	114.		x	[105]
	Megaptera novaeangliae	Humpback whale	77.0		x	[105]

	Megaptera novaeangliae	Humpback whale	95.0		x	[192]
Delphinidae	Cephalorhynchus commersonii	Southern dolphin	18.0		x	[94] [186]
	Delphinus delphis	Common dolphin (+)	20.0		x	[175] [240] [284]
	Globicephala sieboldii (= macrorhynchus)	Pilot whale	63.0		f	[149]
	Globicephala sieboldii (= macrorhynchus)	Pilot whale	46.0		m	[149]
	Grampus griseus	Risso's dolphin	20.0		x	[229]
	Lagenorhynchus acutus	Atlantic white-sided dolphin	27.0		f	[10] [212] [281] [284] [333]
	Lagenorhynchus acutus	Atlantic white-sided dolphin	22.0		m	[10] [212] [281] [284] [333]
	Lagenorhynchus obliquidens	Pacific white-sided dolphin		20.0	x	[10] [212] [281] [284] [333]
	Orcinus orca	Killer whale	90.0		x	[21]
	Orcinus orca	Killer whale	50.0		x	[105]
	Orcinus orca	Killer whale	100.		x	[192]
	Peponocephala electra	Many-toothed blackfish	47.0		x	[31] [240]
	Pseudorca crassidens	False killer whale	22.0		x	[240] [247]
	Stenella attenuata	Spotted dolphin	46.0		x	[229]
	Stenella coeruleoalba	Striped dolphin	50.0		x	[229]
	Steno bredanensis	Rough-toothed dolphin	32.0		x	[240]
	Tursiops aduncus (= truncatus)	Bottle-nosed dolphin	35.0		x	[212] [230] [270] [283] [285]
	Tursiops truncatus	Bottle-nosed dolphin	25.0		x	[21]
	Tursiops truncatus	Bottle-nosed dolphin (+)	25.0	30.0	x	[100]
	Tursiops truncatus	Bottle-nosed dolphin	25.0		x	[212] [230] [270] [283] [285] [21]
Eschrichtidae	Eschrichtius robustus	Gray whale	70.0		x	[105]
	Eschrichtius robustus	Gray whale	77.0		x	[192]
Iniidae	Inia geoffrensis	Amazon dolphin	30.0	18.0	x	[30]
Monodontidae	Delphinapterus leucas	White, or beluga whale	17.3		x	[21]
	Delphinapterus leucas	White, or beluga whale	30.0		x	[25] [27] [282]
	Delphinapterus leucas	White, or beluga whale	40.0		x	[192]
	Monodon monoceros	Narwhal	40.0		x	[18] [65] [194] [255] [305] [192]
Phocoenidae	Neophocaena phocaenoides	Finless porpoise	23.0		x	[148] [192]
	Phocoena phocoena	Common porpoise	13.0		x	[84-86] [290] [192]
	Phocoenoides dalli	Dall's porpoise	22.0		x	[140] [146]
	Phocoenoides dalli	Dall's porpoise (+)	17.0		x	[192]
Physeteridae	Kogia breviceps	Pygmy sperm whale	17.0		x	[192]

	Physeter catodon	Sperm whale	77.0		x	[18] [36] [80] [105] [231, 232] [256]
	Physeter macrocephalus	Sperm whale (+)	32.0		x	[21]
	Physeter macrocephalus	Sperm whale	70.0		x	[192]
Platanistidae	Platanista gangetica	Ganges and Indus dolphins	28.0		m	[147]
	Platanista gangetica	Ganges dolphin (+)	28.0		x	[192]
	Platanista minor	Indus dolphin (+)	28.0		x	[192]
Pontoporiidae	Pontoporia blainvillei	La Plata dolphin	16.0		x	[30] [107] [108] [150] [192]
Ziphiidae	Berardius bairdii	Baird's beaked whale	71.0		x	[199]
	Hyperoodon ampullatus	Northern bottle-nosed whale	37.0		x	[14] [192]
	Mesoplodon densirostris	Beaked whale	27.0		f	[199]
	Ziphius cavirostris	Goose-beaked whale	36.0		x	[199]
Chiroptera						
Hipposideridae	Hipposideros fulvus	Old World leaf-nosed bat	12.0		x	[313]
Molossidae	Tadarida brasiliensis	Free-tailed bat (+)	8.0		x	[174]
Noctilionidae	Noctilio leporinus	Bulldog bat		11.5	x	[143]
Phyllostomidae	Anoura geoffroyi	Geoffroy's long-nosed bat (+)		10.0	x	[143]
	Artibeus jamaicensis	Neotropical fruit bat (+)		10.0	x	[143]
	Carollia perspicillata	Short-tailed leaf-nosed bat (+)		12.4	x	[143]
	Desmodus rotundus	Common vampire bat		19.5	f	[187]
	Glossophaga soricina	Long-tongued nector bat (+)		10.0	x	[143]
	Macrotus californicus		10.4		x	[314]
	Vampyrops lineatus	White-lined bat		10.2	x	[143]
Pteropodidae	Eidolon helvum	Straw-colored fruit bat		21.8	x	[143]
	Pteropus giganteus	Flying fox		31.4	x	[143]
	Rousettus aegypticus.	Rousette fruit bat		22.9	x	[143]
Rhinolophidae	Rhinolophus ferrumequinum	Horseshoe bat (+)	30.0		x	[39]
	Rhinolophus ferrumquinum	Greater horseshoe bat	23.0		x	[100]
	Rhinolophus hipposideros	Lesser horseshoe bat	18.0		x	[100]
Vespertilionidae	Antrozous pallidus	Pallid bat	9.1		x	[182]
	Barbastella barbastellus	Barbastelle	21.0		x	[100]
	Eptesicus fuscus	Big brown bat	19.0		x	[119] [278]
	Eptesicus serotinus	Serotine	6.0		x	[83]
	Lasionycteris noctivagans	Silver-haired bat	12.0		x	[313]
	Lasiurus cinereus	Hoary bat	2.1		m	[313]
	Miniopterus schreibersi	Schreiber's bat	14.0		x	[100]
	Myotis bechsteini	Bechstein's bat	19.0		x	[100]
	Myotis dasyeneme	Pond bat	15.0		x	[100]
	Myotis daubentoni	Daubenton's bat	15.0		x	[100]
	Myotis emarginatus	Geoffroy's bat	15.0		x	[100]
	Myotis evotis		2.2		m	[313]
	Myotis grisescens	Gray myotis	16.5		f	[313]
	Myotis lucifugus	Little brown bat (+)		30.0	x	[151]
	Myotis myotis	Greater mouse-eared bat	18.0		x	[100]
	Myotis mystacinus	Whiskered bat	18.0		x	[100]
	Myotis nattereri	Natterer's bat	15.0		x	[100]
	Myotis nigricans	Black myotis	7.0		x	[313]

	Myotis sodalis	Social bat	13.8		m	[313]
	Myotis thysanodes	Fringed myotis	18.3		m	[313]
	Myotis vivesi	Fish-eating bat	10.0	10.0	x	[313]
	Myotis volans	Long-legged myotis	2.1		m	[313]
	Nyctalus noctula	Noctule	8.0		x	[100]
	Nycticeius humeralis	Evening bat (+)		5.0	x	[328]
	Pipistrellus pipistrellus	Common pipistrelle	9.0		x	[100]
	Pipistrellus subflavus	Pipistrelle	14.8		x	[324]
	Plecotus auritus	Common long-eared bat	12.0		x	[100]
	Plecotus austriacus	Gray long-eared bat	15.0		x	[100]
	Plecotus rafinesquii	Eastern lump-nosed bat	10.1		f	[314]
	Plecotus townsendii	Long-eared bat	10.0		x	[62]
Dermoptera						
Cynocephalidae	Cynocephallus spp.	Colugo (flying lemur)		17.5	x	[176]
Hyraccoidea						
Procaviidae	Dendrohyrax arboreus	Tree hyrax (+)	10.0		x	[100]
	Dendrohyrax arboreus	Tree hyrax		12.3	x	[143]
	Dendrohyrax dorsalis	Tree hyrax		5.5	x	[143]
	Heterohyrax brucei	Bush hyrax	14.0		x	[100]
	Heterohyrax spp.	Gray hyrax	11.0		f	[120-124]
	Procavia capensis	Rock hyrax	14.0		x	[100]
	Procavia capensis	Rock hyrax	8.5	11.0	x	[123] [143]
Insectivora						
Erinaceidae	Echinosorex gymnurus	Moon rat		4.6	x	[100]
	Echinosorex gymnurus	Moon rat		4.4	x	[143]
	Erinaceus algirus	Algerian hedgehog		7.0	x	[100]
	Erinaceus europaeus	Hedgehog	14.0		x	[21]
	Erinaceus europaeus	Eurasian hedgehog	7.0	7.0	x	[45]
	Erinaceus europaeus	West European hedgehog	8.0		x	[100]
	Hemiechinus auritus	Long-eared desert hedgehog		6.8	x	[143]
	Paraechinus aethiopicus	Desert hedgehog		4.5	x	[143]
Solenodontidae	Solenodon cubanus	Cuban solenodon		6.0	x	[100]
	Solenodon paradoxus	Haitian solenodon		11.3	x	[100] [229]
Soricidae	Blarina brevicauda	Short-tailed shrew	2.5		x	[100]
	Crocidura leucodon	Bicolored white-toothed shrew	3.0		x	[100]
	Crocidura russula	Common European white-toothed shrew	3.2		x	[100]
	Crocidura suaveolens	Lesser white-toothed shrew		2.7	x	[100]
	Crocidura spp.	White-toothed shrew		4.0	x	[229]
	Myosorex varius	Forest shrew	1.0		x	[100]
	Neomys anomalus	Southern water shrew	2.0		x	[100]
	Neomys fodiens	Eurasian water shrew	3.0		x	[100]
	Neomys fodiens	Eurasian water shrew		1.6	x	[209]
	Sorex alpinus	Alpine shrew	1.3		x	[100]
	Sorex araneus	Eurasian common shrew	2.0		x	[100]
	Sorex caecutiens	Laxmann's shrew	2.0		x	[100]
	Sorex fumeus	Smoky shrew	1.0		x	[100]
	Sorex minutus	Lesser shrew	2.0		x	[100]
	Sorex trowbridgei	Trowbridge's shrew	1.5		x	[100]

	Sorex vagrans	Vagrant shrew	1.3		x	[100]
	Sorex spp.	Long-tailed shrew		2.0	f	[229]
	Suncus etruscus	Pygmy white-toothed shrew	2.0		x	[21]
	Suncus etruscus	Savi's pygmy shrew	1.3	2.7	x	[100]
	Suncus murinus	House shrew	2.5		x	[100]
	Suncus spp.	Musk shrew		2.5	x	[207]
	Tupaia belangeri	Tree shrew		10.0	x	[100]
	Tupaia glis	Tree shrew (+)	12.0		x	[100]
	Cryptotis spp.	Small-eared shrew		2.6	x	[214]
Talpidae	Galemys pyrenaicus	Pyrenean desman	4.0		x	[100]
	Galemys pyrenaicus	Pyrenean desman	3.5		x	[258]
	Parascalops breweri	Hairy-tailed mole	4.0		x	[105]
	Parascalops breweri	Hairy-tailed mole	5.0		x	[100]
	Scalopus aquaticus	Eastern aquatic mole	3.0		x	[100] [109]
	Talpa europaea	European mole, common mole	4.0		x	[100]
Talpidae	Talpa europaea	European mole	3.0		x	[192]
	Talpa micrura	Eastern mole	3.0		x	[100]
Tenrecidae	Echinops telfairi	Lesser Madagascar "hedgehog"	1.2	13.0	x	[100] [143]
	Hemicentetes nigriceps	Black-headed streaked tenrec	2.6		x	[100]
	Hemicentetes semispinosus	Streaked tenrec		2.6	x	[61] [63]
	Hemicentetes semispinosus	Streaked tenrec	2.7		x	[100]
	Microgale dobsoni	Dobson's long-tailed tenrec		5.6	x	[64]
	Microgale dobsoni	Dobson's long-tailed tenrec	5.6		x	[100]
	Microgale talazaci	Talazaci's long-tailed tenrec		5.8	x	[64]
	Microgale talazaci	Talazaci's long-tailed tenrec	5.8		x	[100]
	Setifer setosus	Large Madagascar "hedgehog"		4.6	x	[100]
	Setifer setosus	Large Madagascar "hedgehog"		10.5	x	[143]
	Tenrec ecaudatus	Tailless tenrec		5.0	x	[100]
	Tenrec ecaudatus	Tenrec		6.3	x	[143]
Lagomorpha						
Leporidae	Lepus americanus	Snowshoe rabbit	5.0		x	[92]
	Lepus californicus	Black-tailed jack rabbit		6.8	x	[60]
	Lepus europaeus	European hare	12.0		x	[100]
	Lepus europaeus	European hare		7.4	x	[143]
	Lepus timidus	Mountain hare	9.0		x	[100]
	Oryctolagus cuniculus	Old World rabbit	10.0		x	[100]
	Oryctolagus cuniculus	Old World rabbit	9.0		x	[101]
	Oryctolagus cuniculus	Old World rabbit (+)	18.0		x	[21]
	Sylvilagus floridanus	Eastern cottontail (+)	5.0	9.0	x	[280]
Ochotonidae	Ochotona collaris	Collared pika	6.0		x	[100]
	Ochotona princeps	N. American pika	6.0		x	[100]
	Ochotona princeps	N. American pika	7.0		x	[192] [210]
	Ochotona pusilla	Small pika	4.0		x	[100]
Macroscelidea						
Macroscelididae	Elephantulus rozetti	N. African elephant shrew (+)		3.0	x	[100]
	Elephantulus rufescens	Rufous elephant shrew	6.0	1.6	x	[100]
	Elephantulus intufi	Long-eared elephant shrew		8.8	x	[143]
	Elephantulus myurus	Long-eared elephant shrew	1.1		f	[297]
	Rhynchocyon chrysopygus	Golden-rumped elephant shrew	4.0		x	[100]

Marsupialia						
Burramyidae	Acrobates pygmaeus	Pygmy gliding possum		7.2	x	[143]
	Acrobates pygmaeus	Pygmy gliding possum (+)	4.0		x	[100]
	Burramys parvus	Mountain pygmy possum	4.0	7.0	x	[229] [100]
	Cercartetus nanus	"Dormouse" possum		8.0	x	[44]
	Distoechurus pennatus	Feather-tailed possum		1.5	x	[44] [332] [336]
Dasyuridae	Antechinomys laniger	Kultarr		3.0	x	[44]
	Antechinus flavipes	Yellow-footed marsupial mouse	1.0		m	[100]
	Antechinus flavipes	Yellow-footed marsupial mouse	2.0		f	[100]
	Antechinus spp.	Broad-footed marsupial mouse (+)		3.0	f	[44]
	Antechinus spp.	Broad-footed marsupial mouse		2.7	m	[261]
	Antechinus stuartii	Brown antechinus	0.9		m	[192] [261]
	Antechinus stuartii	Brown antechinus	3.0		f	[192] [261]
	Dasycercus cristicauda	Mulgara (+)	6.0		x	[100]
	Dasycercus cristicauda	Mulgara (+)		6.0	x	[304]
	Dasykaluta rosamondae			3.0	f	[304]
	Dasyuroides byrnei	Kowari	4.0	6.3	x	[143]
	Dasyurus albopunctatus	Quoll		3.0	x	[143]
	Dasyurus geoffroii	Quoll		3.0	x	[7]
	Dasyurus maculatus	Quoll		4.0	x	[44]
	Dasyurus viverrinus	Quoll		6.8	x	[143]
	Dasyurus viverrinus	Quoll	5.0		x	[100]
	Ningaui timealeyi	Pilbara ningaui	2.0		x	[100]
	Ningaui spp.	Ningaui		2.1	f	[229]
	Parantechinus apicalis	Speckled marsupial mouse	3.0		x	[100]
	Parantechinus bilarni	Sandstone marsupial mouse	3.0		x	[100]
	Phascogale tapoatafa	Tuan (+)		3.0	m	[48] [304]
	Pseudoantechinus macdonnellensis	Fat-tailed marsupial mouse (+)	1.0		x	[100]
	Sarcophilus harrisii	Tasmanian devil	8.0		x	[100]
	Sarcophilus harrisii	Tasmanian devil		8.2	x	[143]
	Sarcophilus harrisii	Tasmanian devil	7.0		x	[192]
	Sminthopsis crassicaudata	Fat-tailed dunnart	1.5		f	[219]
	Sminthopsis crassicaudata	Fat-tailed dunnart	1.3		m	[219]
	Sminthopsis macroura	Stripe-faced dunnart		4.9	x	[143]
Didelphidae	Caluromys laniger	Woolly opossum		5.0	x	[100]
	Caluromys philander	Woolly opossum		6.3	x	[143]
	Caluromysiops irrupta	Black-shouldered opossum		7.8	x	[44]
	Chironectes minimus	Water opossum (yapok)		2.9	x	[44]
	Didelphis spp.	Large American opossum	3.0	5.0	x	[44] [126] [191] [280] [314]
	Didelphis spp.	Large American opossum	7.0	2.0	x	[100]
	Lutreolina crassicaudata	Thick-tailed opossum		3.0	x	[44]
	Marmosa mexicana	Murine opossum		7.0	x	[100]
	Marmosa spp.	Murine (mouse) opossum (+)	1.0	3.0	x	[126]

	Metachirus nudicaudatus	Brown "four-eyed" opossum		4.0	x	[126]
	Monodelphis domestica	Shorttailed opossum		6.0	x	[100]
	Philander mcilhennyi	Black "four-eyed" opossum		2.3	x	[44]
	Philander opossum	Gray "four-eyed" opossum		2.3	x	[44]
Macropodidae	Bettongia gaimardi	Short-nosed "rat"-kangaroo		11.8	x	[143]
	Caloprymnus campestris	Desert "rat"-kangaroo		13.0	x	[143]
	Dendrolagus dorianus	Doria's tree kangaroo	8.0		x	[100]
	Dendrolagus goodfellow	Goodfellow's tree kangaroo	8.0		x	[100]
	Dendrolagus inustus	Grizzled tree kangaroo	10.0		x	[100]
	Dendrolagus matschiei	Matschie's tree kangaroo		14.0	x	[44]
	Dendrolagus matschiei	Matschie's tree kangaroo	8.0		x	[100]
	Dendrolagus ursinus	Black tree kangaroo		20.2	x	[44]
	Dendrolagus ursinus	Black tree kangaroo	8.0		x	[100]
	Dorcopsis veterum	New Guinean forest wallaby		7.6	x	[143]
	Dorcopsis spp.	Forest wallaby (+)	8.0		x	[100]
	Lagostrophus fasciatus	Banded hare-wallaby (+)	4.0		x	[100]
	Macropus agilis	Agile wallaby		10.2	x	[44] [143] [162]
	Macropus agilis	Agile wallaby (+)	10.0		x	[100]
	Macropus antilopinus	Antilopine wallaroo		15.9	x	[44] [143] [162]
	Macropus antilopinus	Antilopine wallaroo (+)	16.0		x	[100]
	Macropus dorsalis	Black-striped wallaby		12.4	x	[44] [143] [162]
	Macropus dorsalis	Black-striped scrub wallaby	15.0		x	[100]
	Macropus eugenii	Tammar wallaby		9.8	x	[44] [143] [162]
	Macropus eugenii	Tammar wallaby	14.0		x	[100]
	Macropus fuliginosus	Western gray kangaroo	20.0		x	[100]
	Macropus giganteus	Eastern gray kangaroo	20.0		x	[100]
	Macropus giganteus	Eastern gray kangaroo	19.8	24.0	x	[162] [196]
	Macropus parma	Parma wallaby	8.0		x	[100]
	Macropus parma	Parma wallaby	7.0		x	[162] [196]
	Macropus parryi	Whiptail wallaby		9.7	x	[44] [143] [162]
	Macropus parryi	Pretty-faced wallaby (+)	10.0		x	[100]
	Macropus robustus	Wallaroo	18.5	19.6	x	[162] [196] [100]
	Macropus rufogriseus	Red-necked wallaby	15.0		x	[100]
	Macropus rufogriseus	Red-necked wallaby	18.5	15.2	x	[162] [196]
	Macropus rufus	Red kangaroo	30.0		x	[21]
	Macropus rufus	Red kangaroo	20.0		x	[100]
	Macropus rufus	Red kangaroo	22.0	16.3	x	[162] [196]
	Onychogalea spp.	Nailtail wallaby (+)	7.0		x	[100]
	Onychogalea spp.	Nailtail wallaby		7.3	x	[143]
	Petrogale penicillata	Rock wallaby		14.4	x	[44]
	Petrogale spp.	Rock wallaby	14.4		x	[100]
	Potorous tridactyles	Potoroo	7.3	12.0	x	[103] [304]
	Setonix brachyurus	Quokka		5.0	x	[100]
	Setonix brachyurus	Quokka (+)	10.0	7.0	f	[287]
	Setonix brachyurus	Quokka (+)	10.0		m	[287]

	Thylogale billardierii	Tasmanian red-bellied pademelon		8.8	x	[44] [100]
	Wallabia bicolor	Swamp wallaby	15.0	12.4	x	[44]
	Wallabia bicolor	Swamp wallaby (+)	12.0		x	[100]
Microbiotheriidae	Dromiciops australis	Monito del monte		2.2	f	[248]
Myrmecobiidae	Myrmecobius fasciatus	Banded anteater (+)		5.0	m	[303]
	Myrmecobius fasciatus	Banded anteater (+)		6.0	f	[303]
Notoryctidae	Notorytes typhlops	Marsupial mole	1.5		x	[100]
Peramelidae	Echymipera spp.	New Guinean spiny bandicoot		2.8	x	[143]
	Isoodon macrourus	Brindled bandicoot	3.0		x	[192]
	Isoodon macrourus	Short-nosed bandicoot	1.0		x	[100]
	Isoodon spp.	Short-nosed bandicoot	4.0		x	[100]
	Isoodon spp.	Short-nosed bandicoot		3.0	x	[301]
	Perameles gunnii	Long-nosed bandicoot		3.0	x	[44]
	Perameles spp.	Long-nosed bandicoot	3.0		x	[44]
	Peroryctes raffrayanus	New Guinean bandicoot		3.3	x	[143]
	Echymipera rufescens	New Guinean spiny bandicoot		2.9	x	[100]
Petauridae	Dactylopsila spp.	Striped possum		4.9	x	[143]
	Dactylopsila trivirgata	Striped possum (+)	5.0		x	[100]
	Gymnobelideus leadbeateri	Leadbeater's possum		9.0	f	[295]
	Gymnobelideus leadbeateri	Leadbeater's possum	7.5		m	[295]
	Petauroides volans	Greater gliding possum		15.0	x	[44]
	Petauroides volans	Greater glider (+)	6.0		x	[100]
	Petaurus australis	Fluffy glider		10.0	x	[44] [143]
	Petaurus australis	Fluffy glider (+)	10.0		x	[100]
	Petaurus breviceps	Sugar glider		14.0	x	[44] [143]
	Petaurus breviceps	Sugar glider	14.0		x	[100]
	Petaurus norfolcensis	Squirrel glider		11.9	x	[44] [143]
	Pseudocheirus peregrinus	Ring-tailed possum	5.0	8.0	x	[44]
Phalangeridae	Spilocuscus maculatus	Spotted cuscus	11.0		x	[100]
	Spilocuscus spp.	Spotted cuscus (+)		11.0	x	[304]
	Trichosurus caninus	Mountain brush-tailed possum	17.0		x	[304]
	Trichosurus caninus	Mountain brush-tailed possum	12.0		m	[100]
	Trichosurus caninus	Mountain brush-tailed possum	17.0		f	[100]
	Trichosurus vulpecula	Brush-tailed possum	13.0		x	[100]
	Trichosurus vulpecula	Brush-tailed possum	13.0	14.7	x	[143] [304]
	Wyulda squamicaudata	Scaly-tailed possum		4.3	x	[44]
	Wyulda squamicaudata	Scaly-tailed possum (+)		6.0	x	[100]
Phascolarctidae	Phascolarctos cinereus	Koala	17.0	20.0	x	[143]
	Phascolarctos cinereus	Koala	13.0	18.0	x	[192]
	Phascolarctos cinereus	Koala (+)		15.0	x	[100]
Potoroidae	Aepyprymnus rufescens	Rufous "rat"-kangaroo (+)	6.0		x	[100]
	Aepyprymnus rufescens	Rufous "rat"-kangaroo (+)		8.0	x	[44]
	Bettongia spp.	Bettong	6.0		x	[100]
	Potorous spp.	Potoroo	12.0	7.0	x	[100]
Pseudocheiridae	Hemibelideus lemuroides	Lemuroid ringtail possum (+)	4.0		x	[100]
	Pseudocheirus peregrinus	Common ringtail possum	6.0		x	[100]
Tarsipedidae	Tarsipes rostratus	Honey possum	2.0		x	[192]
	Tarsipes spenserae	Honey possum (+)	1.0		x	[100]
Thylacinidae	Thylacinus cynocephalus	Tasmanian wolf		8.4	x	[44]
	Thylacinus cynocephalus	Tasmanian wolf		13.0	x	[100]

Thylacomyidae	Macrotis lagotis	Rabbit-bandicoot		5.0	x	[100]
	Macrotis lagotis	Rabbit-bandicoot		7.2	x	[44]
Vombatidae	Lasiorhinus latifrons	Southern hairy-nosed wombat		20.0	x	[100]
	Lasiorhinus latifrons	Hairy-nosed wombat		24.5	x	[143]
	Lasiorhinus latifrons	Southern hairy-nosed wombat		18.0	x	[192]
	Vombatus ursinus	Common wombat	26.0		x	[21]
	Vombatus ursinus	Common wombat		26.1	x	[44]
	Vombatus ursinus	Common wombat	5.0	26.0	x	[192] [100]
Monotremata						
Ornithorhynchidae	Ornithorhynchus anatinus	Duck-billed platypus	17.0		x	[21]
	Ornithorhynchus anatinus	Duck-billed platypus		17.0	x	[101]
	Ornithorhynchus anatinus	Duck-billed platypus		17.0	x	[100]
	Ornithorhynchus anatinus	Duck-billed platypus		17.0	x	[143]
Tachyglosidae	Tachyglossus aculeatus	Short-nosed echidna (+)		50.0	x	[96]
	Tachyglossus aculeatus	Short-nosed echidna		49.4	x	[100]
	Zaglossus bruijni	Long-nosed echidna		30.7	x	[96]
	Zaglossus bruijni	Long-nosed echidna		31.0	x	[100]
Perissodactyla						
Equidae	Equus asinus	African wild ass	20.0		x	[100]
	Equus asinus	African wild ass		47.0	x	[331]
	Equus burchelli (= quagga)	Burchell's zebra		40.0	x	[99] [160]
	Equus caballus	Horse		50.0	x	[15] [326] [331]
	Equus caballus	Horse	62.0		x	[21]
	Equus grevyi	Grevy's zebra	20.0		x	[100]
	Equus grevyi	Grevy's zebra		22.0	x	[160] [163] [229]
	Equus hemionus	Onager	20.0		x	[100]
	Equus hemionus	Onager		38.8	x	[143]
	Equus quagga	Plains zebra (Quagga)	20.0		x	[100]
	Equus zebra	Mountain zebra	20.0		x	[100]
	Equus zebra	Mountain zebra	24.0		f	[239]
Rhinocerotidae	Ceratotherium simum	White rhinoceros	36.0		f	[98] [160] [234, 235] [297]
	Ceratotherium simum	White rhinoceros	50.0		x	[98] [160] [234, 235] [297]
	Ceratotherium simum	White rhinoceros	45.0		x	[192] [100]
	Dicerorhinus sumatrensis	Sumatran rhinoceros	35.0		x	[100]
	Dicerorhinus sumatrensis	Sumatran rhinoceros		32.7	x	[143]
	Dicerorhinus sumatrensis	Sumatran rhinoceros	32.0		x	[192]
	Diceros bicornis	Black rhinoceros (+)		45.0	x	[143]
	Diceros bicornis	Black rhinoceros	40.0		x	[192] [100]
	Rhinoceros sondaicu	Javan rhinoceros	40.0		x	[100]
	Rhinoceros sondaicus	Javan rhinoceros		21.0	x	[170] [172]
	Rhinoceros unicornis	Indian rhinoceros	49.0		x	[21]
	Rhinoceros unicornis	Indian rhinoceros	40.0		x	[100]
	Rhinoceros unicornis	Greater Indian rhinoceros		47.0	x	[172]
	Rhinoceros unicornis	Indian rhinoceros	45.0		x	[192]
Tapiridae	Tapirus indicus	Malayan tapir	30.0		x	[192]

	Tapirus terrestris	Brazilian tapir	30.0		x	[192] [100]
	Tapirus terrestris	Brazilian tapir	35.0		x	[21]
	Tapirus terrestris	Brazilian tapir		35.0	x	[143]
Pholidota						
Manidae	Manis crassicaudata	Pangolin		13.5	x	[143]
	Manis crassicaudata	Pangolin		13.0	x	[192]
Pinnipedia						
Odobenidae	Odobenus rosmarus	Walrus	16.8		x	[21]
	Odobenus rosmarus	Walrus (+)	40.0		x	[68-70] [192]
	Odobenus rosmarus	Walrus	40.0		x	[192]
	Odobenus rosmarus	Walrus	40.0		x	[100]
Otariidae	Arctocephalus galapagoensis	Southern fur seal	22.0		f	[42]
	Arctocephalus gazella	Antarctic fur seal (+)	13.0		m	[192]
	Arctocephalus gazella	Antarctic fur seal	23.0		f	[192]
	Arctocephalus pusillus	Southern fur seal	18.0		x	[157] [327]
	Callorhinus ursinus	Northern fur seal	30.0		x	[9]
	Callorhinus ursinus	Northern fur seal	25.0		x	[192]
	Eumetopias jubatus	Northern sea lion	23.0		x	[100]
	Eumetopias jubatus	Northern sea lion	30.0		f	[157] [189] [279]
	Otaria byronia	S. American sea lion	20.0		x	[100]
	Otaria flavescens	Southern sea lion		24.8	x	[143]
	Zalophus californianus	California sea lion	17.0		x	[100]
	Zalophus californianus	California sea lion		30.0	x	[157]
Phocidae	Cystophora cristata	Hooded seal	35.0		x	[23]
	Cystophora cristata	Hooded seal	34.0		m	[192]
	Cystophora cristata	Hooded seal (+)	35.0		f	[192]
	Erignathus barbatus	Bearded seal	31.0		x	[33] [157]
	Erignathus barbatus	Bearded seal	25.0		m	[192]
	Erignathus barbatus	Bearded seal	31.0		f	[192]
	Halichoerus grypus	Gray seal (+)	46.0		x	[21] [198]
	Halichoerus grypus	Gray seal (+)	40.0		x	[100]
	Halichoerus grypus	Gray seal	31.0	43.0	m	[192]
	Halichoerus grypus	Gray seal	46.0		f	[192]
	Hydrurga leptonyx	Leopard seal	26.0		x	[157] [166] [291] [100]
	Hydrurga leptonyx	Leopard seal (+)	23.0		m	[192]
	Hydrurga leptonyx	Leopard seal (+)	26.0		f	[192]
	Leptonychotes weddelli	Weddell seal	21.0		m	[192]
	Leptonychotes weddelli	Weddell seal	25.0		f	[192]
	Leptonychotes weddelli	Weddell seal	25.0		x	[198] [100]
	Lobodon carcinophagus	Crabeater seal	23.0		x	[100]
	Lobodon carcinophagus	Crabeater seal	39.0		x	[157] [165]
	Lobodon carcinophagus	Crabeater seal	36.0		f	[192]
	Lobodon carcinophagus	Crabeater seal (+)	30.0		m	[192]
	Mirounga angustirostris	Northern elephant seal	14.0	15.0	x	[143]
	Mirounga angustirostris	Northern elephant seal	14.0		m	[192]
	Mirounga angustirostris	Northern elephant seal	14.0		f	[192]
	Mirounga leonina	Southern elephant seal	20.0		x	[100]
	Mirounga leonina	Southern elephant seal	20.0		m	[192]
	Mirounga leonina	Southern elephant seal	18.0		f	[192]

	Mirounga leonina	Southern elephant seal	23.0		x	[118]
	Monachus monachus	Mediterranean monk seal		23.7	x	[143]
	Monachus schauinslandi	Hawaiian monk seal	30.0		x	[154] [157] [192]
	Ommatophoca rossi	Ross seal	21.0		x	[157] [254]
	Ommatophoca rossi	Ross seal	21.0		m	[192]
	Ommatophoca rossi	Ross seal	19.0		f	[192]
	Pagophilus groenlandicus	Harp seal (+)	30.0		x	[100]
	Phoca caspica	Caspian seal	35.0		x	[100]
	Phoca caspica	Caspian seal	50.0		x	[157]
	Phoca caspica	Caspian seal	47.0		m	[192]
	Phoca caspica	Caspian seal	50.0		f	[192]
	Phoca fasciata	Ribbon seal	30.0		x	[34] [157]
	Phoca fasciata	Ribbon seal	31.0		m	[192]
	Phoca fasciata	Ribbon seal	23.0		f	[192]
	Phoca groenlandica	Harp seal (+)	30.0		m	[157] [264]
	Phoca groenlandica	Harp seal (+)	30.0		f	[157] [264] [192]
	Phoca groenlandica	Harp seal	29.0		m	[192]
	Phoca hispida	Ringed seal	46.0		x	[100]
	Phoca hispida	Ringed seal	43.0		m	[192]
	Phoca hispida	Ringed seal	40.0		f	[192]
	Phoca hispida	Ringed seal	43.0		x	[82] [157]
	Phoca largha	Spotted seal	32.0		x	[19] [157] [198]
	Phoca largha	Spotted seal	35.0		x	[100]
	Phoca largha	Spotted seal	29.0		m	[192]
	Phoca largha	Spotted seal	32.0		f	[192]
	Phoca sibirica	Baikal seal	56.0		x	[157] [309]
	Phoca sibirica	Baikal seal	52.0		m	[192]
	Phoca sibirica	Baikal seal	56.0		f	[192]
	Phoca vitulina	Harbor seal	40.0		x	[100]
	Phoca vitulina	Harbor seal	26.0		m	[192]
	Phoca vitulina	Harbor seal	32.0		f	[192]
	Phoca vitulina	Harbor seal	34.0		x	[233]
Primates						
Callimiconidae	Callimico goeldii	Goeldi's monkey (+)		9.0	x	[100]
	Callimico goeldii	Goeldi's monkey		17.9	x	[104]
	Callimico goeldii	Goeldi's monkey		16.1	f	[104]
	Callimico goeldii	Goeldi's monkey (+)		15.8	m	[104]
	Callimico goeldii	Goeldi's marmoset		10.9	x	[143]
Callitrichidae	Callithrix argentata	Bare-eared marmoset		16.8	f	[104]
	Callithrix argentata	Bare-eared marmoset		8.8	x	[104]
	Callithrix humeralifer	Tassel-eared marmoset		15.0	x	[104]
	Callithrix jacchus	Common marmoset (+)	10.0		x	[100]
	Callithrix jacchus	Common marmoset (+)		16.8	m	[104]
	Callithrix jacchus	Common marmoset (+)		15.7	x	[104]
	Callithrix jacchus	Common marmoset		12.0	x	[223]
	Callithrix jacchus	Common marmoset	10.0	16.0	x	[229]
	Cebuella pygmaea	Pygmy marmoset (+)	10.0		x	[100]
	Cebuella pygmaea	Pygmy marmoset (+)		15.1	f	[104]

	Cebuella pygmaea	Pygmy marmoset		11.5	x	[143]
	Leontopithecus rosalia	Golden lion marmoset (+)		24.8	m	[104]
	Leontopithecus rosalia	Golden lion marmoset		22.0	f	[104]
	Leontopithecus spp.	Lion tamarin (+)		15.0	x	[229]
	Saguinus fuscicollis	Saddle-back tamarin		24.5	f	[104]
	Saguinus fuscicollis	Saddle-back tamarin		20.4	f	[104]
	Saguinus imperator	Long-tusked tamarin		20.2	m	[104]
	Saguinus imperator	Long-tusked tamarin		13.0	m	[104]
	Saguinus imperator	Long-tusked tamarin (+)		17.9	x	[143]
	Saguinus nigricollis	Black and red tamarin (+)	10.0		x	[100]
	Saguinus nigricollis	Black and red tamarin		15.2	f	[104]
	Saguinus nigricollis	Black and red tamarin		11.1	x	[104]
	Saguinus oedipus	Cotton top tamarin (+)		23.1	m	[104]
	Saguinus oedipus	Cotton top tamarin (+)		23.0	x	[104]
	Saguinus tamarin			13.2	f	[104]
	Saguinus tamarin			15.4	m	[104]
Cebidae	Alouatta caraya	Black and gold howler		20.3		[104]
	Alouatta caraya	Black and gold howler		20.0		[104]
	Alouatta palliata	Mantled howler (+)	20.0		x	[38] [81] [100] [269]
	Alouatta palliata	Mantled howler (+)	25.0			[104]
	Alouatta seniculus	Red howler	25.0			[269]
	Alouatta villosa	Guatemalan howler (+)	20.0		x	[125]
	Aotus spp.	Night monkey		18.5	x	[143]
	Aotus trivirgatus	Night/owl monkey	12.0		x	[100]
	Aotus trivirgatus	Night/owl monkey		25.3		[104]
	Aotus trivirgatus	Night/owl monkey		19.3		[104]
	Aotus trivirgatus	Night/owl monkey		18.5	x	[143]
	Aotus trivirgatus	Night/owl monkey	20.0			[269]
	Ateles belzebuth	White-bellied spider monkey		28.0		[104]
	Ateles belzebuth	White-bellied spider monkey		26.0	m	[104]
	Ateles belzebuth	White-bellied spider monkey		26.0	f	[104]
	Ateles fusciceps	Brown-headed spider monkey		21.5		[104]
	Ateles fusciceps	Brown-headed spider monkey		24.0		[269]
	Ateles paniscus	Black spider monkey (+)	20.0		x	[100]
	Ateles paniscus	Black spider monkey (+)		37.8	f	[104]
	Ateles paniscus	Black spider monkey (+)		34.0	f	[104]
	Ateles spp.	Spider monkey		33.0	x	[143]
	Cacajao calvus	Bald uakari (+)	18.0		x	[100]
	Cacajao calvus	Bald uakari		22.3	f	[104]
	Cacajao calvus	Bald uakari (+)		19.9	x	[104]
	Cacajao calvus	Bald uakari (+)		20.5	x	[143]
	Cacajao melanocephalus	Black-headed uakari (+)	12.0		x	[100]
	Cacajao melanocephalus	Black-headed uakari (+)		18.0	x	[104]
	Cacajao (calvus) rubicundus	Bald uakari (+)		27.0	m	[104]
	Cacajao (calvus) rubicundus	Bald uakari (+)		23.0	f	[104]
	Callicebus moloch	Dusky titi (+)		25.3	m	[104]
	Callicebus moloch	Dusky titi (+)		20.3	f	[104]
	Callicebus moloch	Dusky titi		12.0	x	[143]
	Cebus albifrons	White-fronted capuchin (+)	40.0		x	[100]
	Cebus albifrons	White-fronted capuchin (+)		44.0	x	[104]

	Cebus albifrons	White-fronted capuchin		40.5	f	[104]
	Cebus albifrons	White-fronted capuchin		25.0	m	[104]
	Cebus apella	Brown capuchin (+)	40.0		x	[100]
	Cebus apella	Brown capuchin		45.1	m	[104]
	Cebus apella	Brown capuchin		41.0	m	[104]
	Cebus apella	Capuchin		44.6	x	[143]
	Cebus capucinus	White-faced capuchin		47.0	x	[100]
	Cebus capucinus	White-faced capuchin		54.8	m	[104]
	Cebus capucinus	White-faced capuchin		46.9	m	[104] [143]
	Cebus capucinus	White-faced capuchin (+)		40.0	x	[142]
	Cebus nigrivittatus	Weeper capuchin (+)	30.0		x	[100]
	Cebus nigrivittatus	Weeper capuchin (+)		41.0	f	[104]
	Chiropotes albinasus	White-nosed saki (+)	17.0		x	[100]
	Chiropotes albinasus	White-nosed saki (+)		11.7	f	[104]
	Chiropotes satanas	Bearded saki (+)	18.0		x	[100]
	Chiropotes satanas	Bearded saki		15.0	f	[104]
	Chiropotes satanas	Bearded saki		12.7	m	[104]
	Chiropotes satanas	Bearded saki		15.0	x	[223]
	Lagothrix lagotricha	Woolly monkey		25.0	x	[100]
	Lagothrix lagotricha	Woolly monkey		30.0	f	[104]
	Lagothrix lagotricha	Woolly monkey		25.9	x	[104]
	Lagothrix lagotricha	Woolly monkey		24.8	m	[143]
	Pithecia aequatorialis	Saki		14.9	x	[143]
	Pithecia monachus	Red-bearded saki (+)		14.0	x	[100]
	Pithecia monachus	Red-bearded saki		24.6	f	[104]
	Pithecia pithecia	White-faced saki (+)	14.0		x	[100]
	Pithecia pithecia	White-faced saki (+)		20.7	m	[104]
	Pithecia pithecia	White-faced saki (+)		18.0	m	[104]
	Pithecia pithecia	White-faced saki (+)		18.0	f	[104]
	Saimiri sciureus	Common squirrel monkey	21.0		x	[100]
	Saimiri sciureus	Common squirrel monkey		27.0	f	[104]
	Saimiri sciureus	Common squirrel monkey		22.7	m	[104]
	Saimiri sciureus	Common squirrel monkey		21.0	x	[223]
	Saimiri spp.	Squirrel monkey (+)		15.3	x	[143]
Cercopithecidae	Allenopithecus nigroviridis	Allen's monkey (+)	23.0		x	[100]
	Cercocebus (= Lophocebus) aterrimus	Black mangabey		26.8	m	[104]
	Cercocebus albigena	Gray-cheeked mangabey	21.0		x	[110]
	Cercocebus albigena	Gray-cheeked mangabey		32.7	x	[143]
	Cercocebus atys	Sooty mangabey		26.8	m	[104]
	Cercocebus atys	Sooty mangabey	18.0		x	[110]
	Cercocebus galeritus	Tana River mangabey (+)		21.0	f	[104]
	Cercocebus galeritus	Tana River mangabey		19.0	x	[269]
	Cercocebus torquatus	Collared mangabey (+)	30.0		x	[100]
	Cercocebus torquatus	Collared mangabey (+)		14.7	f	[104]
	Cercocebus torquatus	Collared mangabey	20.5		x	[110]
	Cercocebus torquatus	Collared mangabey (+)		27.0	x	[269]
	Cercopithecus (= Chlorocebus) aethiops	Vervet (+)	30.0		x	[100]
	Cercopithecus (= Chlorocebus) aethiops	Vervet (+)		23.0	m	[104]

Cercopithecus (= Chlorocebus) aethiops	Vervet (+)		31.6	m	[104]
Cercopithecus (= Chlorocebus) aethiops	Vervet	31.0		x	[110]
Cercopithecus ascanius	Red-tailed guenon		28.3	f	[104]
Cercopithecus ascanius	Red-tailed guenon		25.9	f	[104]
Cercopithecus campbelli	Campbell's guenon (+)		25.0	f	[104]
Cercopithecus campbelli	Guenon (+)		33.0	x	[143]
Cercopithecus cephus	Mustached guenon (+)		23.0	m	[104]
Cercopithecus cephus	Mustached guenon (+)		23.0	f	[104]
Cercopithecus diana	Diana monkey		19.0	x	[100]
Cercopithecus diana	Diana monkey		37.3	f	[104]
Cercopithecus diana	Diana monkey (+)		30.0	f	[104]
Cercopithecus diana	Diana monkey	34.8		x	[110]
Cercopithecus diana	Diana monkey		34.8	x	[142]
Cercopithecus hamlyni	Owl-faced monkey		27.0	m	[104]
Cercopithecus l'hoesti	L'hoest's monkey (+)		16.0	x	[100]
Cercopithecus mitis	Blue monkey (+)		27.1	f	[104]
Cercopithecus mitis	Blue monkey (+)		25.0	f	[104]
Cercopithecus mona	Mona monkey (+)	30.0		x	[100]
Cercopithecus mona	Mona monkey (+)		20.5	f	[104]
Cercopithecus mona	Mona monkey		22.0	x	[269]
Cercopithecus neglectus	De Brazza's monkey		26.3	f	[104]
Cercopithecus neglectus	De Brazza's monkey		23.0	f	[104]
Cercopithecus nictitans	Greater spot-nosed monkey		23.0	f	[104]
Cercopithecus nictitans	Greater spot-nosed monkey		18.9	m	[104]
Cercopithecus petaurista	Lesser spot-nosed guenon (+)		19.0	x	[104]
Cercopithecus petaurista	Lesser spot-nosed guenon (+)		16.7	f	[104]
Cercopithecus pogonias	Crowned guenon		24.1	f	[104]
Cercopithecus pogonias	Crowned guenon		18.5	f	[104]
Colobus guereza	Guereza		24.0	x	[100]
Colobus guereza	Guereza		24.5	m	[104]
Colobus guereza	Guereza		23.8	f	[104]
Colobus polykomos	Western black and white colobus (+)		27.5	x	[115]
Colobus polykomos	Western black and white colobus (+)		23.0	x	[100]
Colobus polykomos	Western black and white colobus		24.0	m	[104]
Colobus polykomos	Western black and white colobus	26.0		x	[110]
Colobus polykomos	Western black and white colobus	30.5		x	[269]
Erythrocebus patas	Patas monkey (+)	21.0		x	[100]
Erythrocebus patas	Patas monkey		23.9	f	[104]
Erythrocebus patas	Patas monkey		21.7	f	[104]
Erythrocebus patas	Patas monkey		21.6	x	[143]
Macaca arctoides	Stump-tailed macaque (+)		24.0	f	[104]
Macaca arctoides	Stump-tailed macaque (+)		20.1	f	[104]
Macaca arctoides	Stump-tailed macaque (+)	30.0		x	[110]
Macaca fascicularis	Crab-eating monkey		38.0	x	[100]

Macaca fascicularis	Crab-eating monkey		25.0	f	[104]
Macaca fascicularis	Crab-eating macaque		37.1	x	[143]
Macaca fuscata	Japanese macaque		32.0	f	[71]
Macaca fuscata	Japanese macaque		27.0	m	[71]
Macaca fuscata	Japanese macaque (+)	30.0		x	[100]
Macaca fuscata	Japanese macaque		33.0	x	[104]
Macaca fuscata	Japanese macaque (+)		22.7	f	[104]
Macaca mulatta	Rhesus macaque (+)	30.0		x	[100]
Macaca mulatta	Rhesus macaque (+)		26.0	x	[104]
Macaca mulatta	Rhesus macaque (+)		23.0	f	[104]
Macaca mulatta	Rhesus macaque		36.0	x	[104]
Macaca mulatta	Rhesus macaque		35.0	x	[104]
Macaca nemestrina	Pig-tailed macaque (+)	30.0		x	[100]
Macaca nemestrina	Pig-tailed macaque (+)		34.3	x	[104]
Macaca nemestrina	Pig-tailed macaque		27.1	m	[104]
Macaca radiata	Bonnet macaque (+)	20.0		x	[100]
Macaca silenus	Lion-tailed macaque (+)	30.0		x	[100]
Macaca silenus	Lion-tailed macaque		40.0	f	[104]
Macaca silenus	Lion-tailed macaque		38.0	m	[104]
Macaca sinica	Toque macaque		35.0	x	[76]
Macaca sinica	Toque macaque		29.3	x	[104]
Macaca sinica	Toque macaque	30.0		x	[269]
Macaca sylvanus	Barbary ape (+)	20.0		x	[100]
Macaca sylvanus	Barbary ape (+)		17.0	m	[104]
Macaca sylvanus	Barbary ape	22.0		x	[269]
Mandrillus leucophaeus	Drill		33.4	f	[104]
Mandrillus leucophaeus	Drill		28.4	x	[104]
Mandrillus leucophaeus	Drill		28.6	x	[110]
Mandrillus sphinx	Mandrill		46.0	x	[100] [116]
Mandrillus sphinx	Mandrill		31.7	f	[104]
Miopithecus talapoin	Talapoin		27.7	x	[143]
Miopithecus talapoin	Talapoin	28.0		x	[100]
Miopithecus talapoin	Dwarf guenon (Talapoin)		30.9	m	[104]
Miopithecus talapoin	Dwarf guenon (Talapoin) (+)		23.1	m	[104]
Miopithecus talapoin	Dwarf guenon (Talapoin)		27.7	x	[143]
Nasalis larvatus	Proboscis monkey		13.6	x	[97]
Nasalis larvatus	Proboscis monkey		21.0	m	[104]
Nasalis larvatus	Proboscis monkey (+)		20.0	f	[104]
Papio anubis	Olive baboon		25.2	m	[104]
Papio cynocephalus	Yellow baboon	45.0		x	[100]
Papio cynocephalus	Yellow baboon (+)		35.1	f	[104]
Papio cynocephalus	Yellow baboon		27.7	x	[104]
Papio cynocephalus	Yellow baboon	40.0		x	[269]
Papio hamadryas	Hamadryas baboon (+)	37.0		x	[100]
Papio hamadryas	Hamadryas baboon		37.5	m	[104] [229]
Papio hamadryas	Hamadryas baboon		35.6	x	[104] [143]
Papio hamadryas	Hamadryas baboon		28.8	x	[104]
Papio papio	Guinea baboon		27.0	m	[104]
Papio papio	Guinea baboon		25.8	x	[104]
Papio papio	Guinea baboon		40.0	x	[269]
Papio ursinus	Chacma baboon (+)		31.2	f	[104]

	Papio ursinus	Chacma baboon		27.6	x	[104]
	Papio ursinus	Chacma baboon		45.0	x	[229]
	Presbytis (= Trachypithecus) obscura	Spectacled leaf monkey(+)		25.0	x	[104]
	Presbytis (= Trachypithecus) obscura	Spectacled leaf monkey(+)		15.3	f	[104]
	Presbytis (= Trachypithecus) cristata	Silvered langur		31.1	f	[104]
	Presbytis (= Trachypithecus) cristata	Silvered langur		18.8	m	[104]
	Presbytis (= Semnopithecus) entellus	Hanuman langur (+)		25.0	f	[100] [104]
	Presbytis (= Semnopithecus) entellus	Hanuman langur (+)		24.0	m	[104]
	Semnopithecus (= Presbytis) entellus	Hanuman langur (+)		25.0	x	[229]
	Presbytis melalophos	Mitered leaf monkey		16.0	x	[104]
	Presbytis senex (= Trachypithecus vetulus)	Purple-faced leaf monkey (+)		23.0	m	[104]
	Pygathrix nemaeus	Red-shanked douc langur (+)		25.0	m	[104]
	Pygathrix nemaeus	Red-shanked douc langur (+)		21.0	m	[104]
	Theropithecus gelada	Gelada (+)	20.0		x	[100]
	Theropithecus gelada	Gelada (+)		28.0	f	[104]
	Theropithecus gelada	Gelada (+)		27.0	f	[104]
	Theropithecus gelada	Gelada (+)		20.8	x	[143]
	Trachypithecus spp.	Leaf monkey		23.7	x	[143]
Cheirogaleidae	Cheirogaleus major	Greater dwarf lemur		15.0	x	[192]
	Cheirogaleus major	Greater dwarf lemur		15.0	x	[100]
	Cheirogaleus major	Greater dwarf lemur		10.0	f	[104]
	Cheirogaleus major	Greater dwarf lemur		8.7	x	[104]
	Cheirogaleus medius	Fat-tailed dwarf lemur		18.0	x	[100]
	Cheirogaleus medius	Fat-tailed dwarf lemur		19.3	m	[104]
	Cheirogaleus medius	Fat-tailed dwarf lemur		17.0	f	[104]
	Cheirogaleus medius	Fat-tailed dwarf lemur		18.0	x	[192]
	Cheirogaleus spp.	Dwarf lemur (+)		8.3	x	[143]
	Microcebus murinus	Gray lesser mouse lemur		14.0	x	[192]
	Microcebus rufus	Brown lesser mouse lemur		12.0	x	[192]
	Mirza coquereli	Coquerel's dwarf lemur		15.3	x	[104] [143]
Daubentoniidae	Daubentonia madagascariensis	Aye-aye		23.3	x	[143]
	Daubentonia madagascariensis	Aye-aye (+)		23.0	x	[100]
	Daubentonia madagascariensis	Aye-aye		24.3	x	[104]
	Daubentonia madagascariensis	Aye-aye (+)		5.4	f	[104]
Hominidae	Homo sapiens	Human (+)		116.0	x	[21]
	Homo sapiens	Human (+)		100.0	x	[229]
	Homo sapiens	Human		122.5	f	[5]
Hylobatidae	Hylobates agilis	Agile gibbon	25.0		x	[100]
	Hylobates agilis	Agile gibbon (+)		44.0	m	[104]
	Hylobates agilis	Agile gibbon (+)		28.0	f	[104]
	Hylobates concolor	Concolor gibbon		44.1	x	[104]
	Hylobates concolor	Concolor gibbon (+)		25.0	f	[104]
	Hylobates lar	Lar gibbon		31.5	x	[142]

	Hylobates lar	Lar gibbon	25.0		x	[192]
	Hylobates lar	Lar gibbon (+)		40.0	m	[104]
	Hylobates lar	Lar gibbon (+)		40.0	f	[104]
	Hylobates moloch muelleri	Gray gibbon	25.0		x	[100]
	Hylobates moloch muelleri	Gray gibbon (+)		29.0	m	[104]
	Hylobates moloch muelleri	Gray gibbon (+)		25.0	f	[104]
	Hylobates pileatus	Capped gibbon	25.0		x	[100]
	Hylobates pileatus	Capped gibbon		36.0	m	[104]
	Hylobates pileatus	Capped gibbon (+)		31.0	m	[104]
	Hylobates pileatus x H. agilis	Gibbon hybrid		37.9	f	[104]
	Hylobates pileatus x H. agilis	Gibbon hybrid		34.7	x	[143]
	Hylobates syndactylus	Siamang	25.0	19.9	x	[41] [100]
	Hylobates syndactylus	Siamang (+)		38.0	f	[104]
	Hylobates syndactylus	Siamang		37.0	m	[104]
	Hylobates syndactylus	Siamang	30.0		x	[192]
Indriidae	Propithecus verreauxi	Sifaka		18.2	x	[143]
	Propithecus verreauxi	Verreaux's sifaka (+)		18.0	x	[100]
	Propithecus verreauxi	Verreaux's sifaka (+)		20.6	m	[104]
	Propithecus verreauxi	Verreaux's sifaka (+)		18.2	f	[104]
Lemuridae	Eulemur macaco X Petterus fulvus	Captive hybrid lemur		39.0	x	[143]
	Hapalemur griseus	Gentle lemur		12.8	m	[104] [100] [143]
	Hapalemur griseus	Gray gentle lemur (+)		17.1	f	[104]
	Hapalemur simus	Broad-nosed gentle lemur		12.0	x	[100]
	Lemur fulvus	Brown lemur		37.0	m	[104]
	Lemur fulvus	Brown lemur		36.5	f	[104]
	Lemur fulvus	Brown lemur		30.8	x	[142]
	Lemur catta	Ring-tailed lemur (+)		20.0	x	[100]
	Lemur catta	Ring-tailed lemur (+)		30.0	f	[104]
	Lemur catta	Ring-tailed lemur (+)		26.0	x	[104]
	Lemur catta	Ring-tailed lemur	27.1		x	[110]
	Lemur coronatus	Crowned lemur		18.4	x	[104]
	Lemur macaco	Black lemur		30.0	x	[100]
	Lemur macaco	Black lemur		28.0	f	[104]
	Lemur macaco	Black lemur	27.1		x	[104]
	Lemur macaco	Black lemur		27.1	f	[223]
	Lemur mongoz	Mongoose lemur		30.0	x	[100]
	Lemur mongoz	Mongoose lemur (+)		27.9	m	[104]
	Lemur mongoz	Mongoose lemur		24.3	x	[104]
	Microcebus coquerrel	Coquerel's mouse lemur		15.0	x	[100]
	Microcebus murinus	Lesser mouse lemur		15.5	m	[104]
	Microcebus murinus	Lesser mouse lemur (+)		14.0	m	[104] [100]
	Phaner furcifer	Fork-marked lemur		12.0	x	[104]
	Varecia variegata	Ruffed lemur		19.0	x	[100] [229]
	Varecia variegata	Ruffed lemur		28.0	m	[104]
	Varecia variegata	Ruffed lemur (+)		32.0	f	[104]
Lorisidae	Arctocebus calabarensis	Golden potto		12.2	x	[104]
	Arctocebus calabarensis	Golden potto	9.5		x	[110]
	Arctocebus calabarensis	Golden potto		13.0	x	[143] [100]
	Galago alleni	Allen's bush baby		12.0	x	[192] [100]

	Galago crassicaudatus	Greater galago		15.0	x	[100]
	Galago crassicaudatus	Greater galago		18.8	m	[104]
	Galago crassicaudatus	Greater galago		14.0	x	[223]
	Galago demidovii	Dwarf bush baby		9.4	x	[104]
	Galago demidovii	Dwarf bush baby		9.2	f	[104]
	Galago demidovii	Dwarf bush baby	14.0		x	[110]
	Galago demidovii	Dwarf bush baby		12.0	x	[192] [100]
	Galago elegantulus	Needle-clawed bush baby		15.0	x	[192] [100]
	Galago senegalensis	Galago (+)		17.0	f	[104]
	Galago senegalensis	Senegal bush baby		16.5	x	[104] [143]
	Galago senegalensis	Lesser bush baby		14.0	x	[192] [100]
	Galagoides alleni	Dwarf galago (+)		8.0	x	[40] [320]
	Loris tardigradus	Slender loris		15.0	x	[100]
	Loris tardigradus	Slender loris (+)		14.0	m	[104]
	Loris tardigradus	Slender loris		16.4	m	[104]
	Loris tardigradus	Slender loris		12.4	x	[143] [229]
	Nycticebus coucang	Slow loris		20.0	f	[104] [100]
	Nycticebus coucang	Slow loris		26.5	f	[104]
	Nycticebus spp	Slow loris	14.0		x	[229]
	Otolemur garnettii	Greater bush baby (+)		14.0	x	[55] [131] [160] [224] [297] [320]
	Otolemur garnettii	Greater bush baby (+)		15.7	m	[104]
	Otolemur garnettii	Greater bush baby		15.0	f	[104]
	Perodicticus potto	Potto (+)		22.0	x	[100]
	Perodicticus potto	Potto		26.0	m	[104]
	Perodicticus potto	Potto		22.0	m	[143]
	Perodicticus potto	Potto		15.0	x	[192]
Megaladapidae	Lepilemur mustilinus	Sportive lemur		12.0	x	[100]
	Lepilemur spp.	Sportive lemur		8.6	x	[143]
Pongidae	Gorilla gorilla	Gorilla	50.0	54.0	x	[229] [21] [104]
	Gorilla gorilla	Gorilla (+)	47.0	40.0	x	[100]
	Gorilla gorilla	Gorilla		47.0	f	[104]
	Gorilla gorilla	Gorilla (+)		39.3	x	[142]
	Gorilla gorilla gorilla	Lowland gorilla		36.0	x	[141]
	Pan paniscus	Pygmy chimpanzee (+)	20.0		x	[100]
	Pan paniscus	Pygmy chimpanzee (+)		48.0	f	[236]
	Pan paniscus	Pygmy chimpanzee (+)		35.0	m	[236]
	Pan troglodytes	Chimpanzee	51.0		x	[21]
	Pan troglodytes	Chimpanzee (+)	50.0	40.0	x	[100]
	Pan troglodytes	Chimpanzee		56.0	m	[104]
	Pan troglodytes	Chimpanzee		59.4	f	[104]
	Pan troglodytes	Chimpanzee (+)		44.5	x	[142]
	Pan troglodytes	Chimpanzee	60.0	53.0	x	[143]
	Pan troglodytes	Common chimpanzee	45.0		x	[192]
	Pongo pygmaeus	Orangutan	59.0		x	[21]
	Pongo pygmaeus	Orangutan	35.0	50.0	x	[192] [100]
	Pongo pygmaeus	Orangutan		59.0	x	[229]
	Pongo pygmaeus	Orangutan		58.8	m	[104]
	Pongo pygmaeus	Orangutan		57.3	f	[104]

Tarsiidae	Tarsius bancanus	Western tarsier	8.0		x	[192] [100]
	Tarsius spectrum	Spectral tarsier		12.0	x	[110]
	Tarsius syrichta	Philippine tarsier (+)		13.0	x	[100]
	Tarsius syrichta	Tarsier		13.4	x	[143]
	Tarsius syrichta	Philippine tarsier		12.0	x	[192]
	Tarsius syrichta	Philippine tarsier		13.0	x	[100]
	Tarsius syrichta	Philippine tarsier		15.0	m	[104]
	Tarsius syrichta	Philippine tarsier		13.5	m	[104]
Proboscidea						
Elephantidae	Elephas maximus	Asiatic elephant	78.0		x	[21]
	Elephas maximus	Asian elephant	40.0	69.0	x	[100]
	Elephas maximus	Asian elephant	60.0	80.0	x	[192]
	Elephas maximus	Asiatic elephant		77.0	x	[288]
	Elephas maximus	Asiatic elephant	70.0		x	[160] [173] [220] [296]
	Loxodonta africana	African elephant	70.0		x	[21] [100]
	Loxodonta africana	African elephant	60.0	80.0	x	[192]
Rodentia						
Abrocomidae	Abrocoma bennetti	Chinchilla-rat		2.3	x	[143]
Anomaluridae	Pedetes capensis	Springhare		13.0	x	[100]
	Pedetes capensis	Springhare		14.5	x	[192]
Aplodontidae	Aplodontia rufa	Mountain beaver	6.0		x	[100]
	Aplodontia rufa	Mountain beaver		6.0	x	[190]
	Aplodontia rufa	Mountain beaver	10.0		x	[192]
Bathyergidae	Cryptomys spp.	Common mole-rat		2.4	x	[143]
	Georychus capensis	Cape mole-rat	3.0		x	[308]
	Heliophobius argenteocinereus	Silvery mole-rat (+)		3.1	f	[136]
	Heterocephalus glaber	Naked mole-rat		5.0	x	[137-139] [135] [100]
	Heterocephalus glaber	Naked mole-rat		10.0	x	[192]
Capromyidae	Capromys pilorides	Cuban hutia		11.3	x	[143]
	Geocapromys (= Capromys) brownii	Brown's hutia		8.3	x	[143]
	Geocapromys (= Capromys) ingrahami	Bahamian hutia (+)	6.0		x	[100]
	Plagiodontia aedium	Hispaniolan hutia		9.9	x	[143]
Castoridae	Castor spp.	Beaver	24.0		x	[29]
	Castor canadensis	American beaver (+)	15.0		x	[21]
	Castor fiber	European beaver	17.0	25.0	x	[100]
Caviidae	Cavia apera f. porcellus	Domestic guinea pig	5.0	10.0	x	[100]
	Cavia porcellus	Guinea pig	14.8		x	[21]
	Cavia spp.	Guinea pig		8.0	x	[229]
	Dolichotis patagonum	Patagonian cavy		14.0	x	[101]
	Galea musteloides	Yellow-toothed cavy	1.3	1.8	x	[168] [265] [329]
	Galea spixii	Yellow-toothed cavy		4.6	x	[143]
	Kerodon rupestris	Rock cavy		11.0	x	[262]
Chinchillidae	Chinchilla chinchilla boliviana	Short-tailed chinchilla	10.0		x	[100]
	Chinchilla lanigera	Chinchilla	11.3		x	[21]
	Chinchilla lanigera	Long-tailed chinchilla	10.0		x	[100]
	Chinchilla spp.	Chinchilla (+)	10.0	20.0	x	[329] [229]

	Lagidium peruanum	Mountain viscacha		19.5	x	[143]
	Lagostomus maximus	Plains viscacha		9.4	x	[229]
	Lagostomus maximus	Viscacha (+)		9.0	x	[100]
Ctenodactylidae	Ctenodactylus spp.	Gundi		5.0	x	[100]
	Ctenodactylus vali	Gundi		5.0	f	[89]
	Pectinator spekei	Speke's pectinator		4.0	f	[89]
	Pectinator spekei	Speke's gundi		10.0	x	[192]
Ctenomyidae	Ctenomys talarum	Tuco-tuco	3.0		x	[329]
	Ctenomys spp.	Tuco-tuco	3.0		x	[100]
Dasyproctidae	Agouti paca	Paca (+)		16.0	x	[43] [229]
	Dasyprocta leporina	Agouti		17.8	x	[143]
	Dasyprocta punctata	Central American agouti (+)	10.0		x	[100]
	Myoprocta spp.	Acouchi		10.2	x	[143]
	Myoprocta acouchy	Acouchi (+)		10.0	x	[100]
Dinomyidae	Dinomys branickii	Pacarana		9.4	x	[143]
	Dinomys branickii	Pacarana (+)		9.0	x	[100]
Dipodidae	Allactaga euphratica	Four- and five-toed jerboa		4.2	x	[143]
Echimyidae	Kannabateomys amblyonyx	Rato de Taquara		1.6	x	[143]
	Makalata armata			3.1	x	[143]
	Proechimys guyannensis	Spiny rat	1.7	3.5	x	[66]
	Proechimys semispinosus	Spiny rat	2.0	4.8	x	[143]
Erethizontidae	Coendou prehensilis	Prehensile-tailed porcupine		17.3	x	[143]
	Coendou prehensilis	Prehensile-tailed porcupine (+)	17.0	9.0	x	[100]
	Erethizon dorsatum	N. American porcupine	18.0		x	[56]
	Erethizon dorsatum	N. American porcupine	17.0		x	[192] [100]
Geomyidae	Cratogeomys castanops	Mexican pocket gopher	2.6		m	[298]
	Cratogeomys castanops	Mexican pocket gopher	4.7		f	[298]
	Geomys bursarius	Eastern pocket gopher		7.2	x	[143]
	Geomys spp.	Eastern pocket gopher	4.0	7.0	x	[100]
	Pappogeomys spp.	Yellow and cinnamon pocket gopher	1.1		f	[100]
	Pappogeomys spp.	Yellow and cinnamon pocket gopher	0.6		m	[100]
	Thomomys spp	Western pocket gopher	5.0		x	[100]
	Thomomys spp.	Western pocket gopher	4.0		x	[229]
	Thomomys talpoides	Pocket gopher	1.6		x	[21]
Gliridae	Dryomys nitedula	Forest dormouse (+)	4.0		x	[100]
	Eliomys quercinus	Garden dormouse		5.5	x	[143]
	Glis glis	Edible dormouse	9.0		x	[100]
	Glis glis	Fat dormouse		8.7	x	[229]
	Graphiurus murinus	African dormouse		5.8	x	[143]
	Muscardinus avellanarius	Common dormouse	4.0	6.0	x	[45] [100]
Heteromyidae	Chaetodipus fallax	Coarse-haired pocket mouse		8.3	x	[143]
	Dipodomys ordii	Kangaroo rat		9.8	x	[299]
	Dipodomys spp.	Kangaroo rat	9.8		x	[100]
	Liomys adspersus	Spiny pocket mouse (+)	1.5		x	[74, 75]
	Liomys spp.	Spiny pocket mouse	1.5		x	[100]
	Microdipodops megacephalus	Kangaroo mouse		5.4	x	[60]
	Perognathus formosus	Hibernating pocket mouse	5.0		x	[192]
	Perognathus longimembris	Silky pocket mouse		8.3	x	[113]
	Perognathus spp.	Pocket mouse	8.0		x	[100]

Hydrochaeridae	Hydrochaerus hydrochaeris	Capybara		12.0	x	[101]
	Hydrochaerus hydrochaeris	Capybara	10.0		x	[192]
	Hydrochaerus hydrochaeris	Capybara	12.0		x	[21]
	Hydrochaerus hydrochaeris	Capybara	10.0		x	[100]
Hystricidae	Atherurus africanus	Brush-tailed porcupine		22.9	x	[299] [100]
	Hystrix brachyura	Old World porcupine		27.3	x	[143]
	Hystrix pumila	Old World porcupine		9.5	x	[143]
	Hystrix spp.	Porcupine (+)	20.0	15.0	x	[100]
	Thecurus (= Hystrix) pumilis	Indonesian porcupine		9.5	x	[100]
	Trichys fasciculata	Long-tailed porcupine		10.1	x	[299] [100]
Muridae	Acomys cahirinus	Spiny mouse		5.0	x	[101] [106] [158] [267]
	Acomys spp.	Spiny mouse		4.0	x	[100]
	Apodemus agrarius	Striped field mouse		4.0	x	[100]
	Apodemus flavicollis	Yellow-necked mouse		4.0	x	[100]
	Apodemus sylvaticus	Wood mouse		4.0	x	[100]
	Apodemus sylvaticus	Old World wood and field mouse	1.0	4.4	x	[143]
	Arvicanthis spp.	Unstriped grass mouse		7.0	x	[100]
	Arvicanthis spp.	Unstriped grass mouse		6.7	x	[143]
	Arvicola richardsoni	Richardson's water vole	1.5		x	[100]
	Arvicola sapidus	Southwestern water vole	2.5	3.5	x	[100]
	Arvicola terrestris	Water vole	2.5	3.5	x	[100]
	Arvicola terrestris	Water vole		5.0	x	[45]
	Baiomys taylori	Pygmy mouse		3.0	x	[143]
	Baiomys taylori	Northern pygmy mouse		3.0	x	[100]
	Berylmys bowersi	White-toothed rat		2.9	x	[207]
	Cannomys badius	Lesser bamboo rat		3.3	x	[143]
	Cannomys spp.	Bamboo rat		4.0	x	[100]
	Chiropodomys spp.	Pencil-tailed tree mouse		4.0	x	[100]
	Chiropodomys spp.	Pencil-tailed tree mouse	2.0	3.6	x	[207] [222]
	Clethrionomys glareolus	Red-backed mouse		4.9	x	[143]
	Colomys goslingi	African water rat		3.0	x	[100]
	Cricetomys spp.	African giant pounched rat		7.8	x	[143]
	Cricetomys spp.	African giant pouched rat (+)		7.0	x	[100]
	Cricetus cricetus	Common hamster	4.0		x	[100]
	Cricetus cricetus	Common hamster		2.0	x	[229]
	Delanymys brooksi	Delany's swamp mouse	2.0		x	[100]
	Dendromus spp.	African climbing mouse		4.0	x	[100]
	Dendromus mesomelas	African climbing mouse		3.3	x	[143]
	Dicrostonyx torquatus	Arctic lemming		3.8	x	[100]
	Dicrostonyx spp.	Collared lemming		3.3	x	[10]
	Eligmodontia typus	Highland desert mouse		0.8	x	[238]
	Eothenomys smithi	Pere David's vole		3.5	x	[6]
	Gerbillus pyramidum	Greater Egyptian gerbil (+)	8.0		x	[21]
	Gerbillus pyramidum	Greater Egyptian gerbil		8.2	x	[143]
	Grammomys dolichurus			4.4	x	[143]
	Hybomys spp.	Back-striped mouse		4.0	x	[100]
	Lemmus lemmus	Norway lemming		2.0	x	[100]
	Lemmus spp.	True lemming		2.0	x	[10] [101] [111]

Lemniscomys striatus	Striped grass mouse		2.5	f	[229]
Lenothrix canus	Gray tree rat		3.8	x	[207]
Leopoldamys spp.	Long-tailed giant rat		2.0	x	[100] [207]
Lophiomys imhausi	Crested rat		7.5	x	[143]
Lophiomys imhausi	Crested rat		1.5	x	[100]
Lophuromys spp.	Brush-furred mouse		2.0	x	[159]
Mastomys spp.	Multimammate mouse		4.0	x	[100]
Meriones crassus	Jird		5.6	x	[143]
Meriones unguiculatus	Mongolian gerbil (+)		2.0	x	[100]
Mesembriomys gouldii	Tree rat		3.9	x	[143]
Mesocricetus auratus	Golden hamster	10.0		x	[21]
Mesocricetus auratus	Golden hamster	3.0		x	[100]
Mesocricetus auratus	Golden hamster		3.0	x	[229]
Micromys minutus	Harvest mouse		4.0	x	[100]
Micromys minutus	Old World harvest mouse	1.5	5.0	x	[229]
Microtus agrestis	Field vole	1.6	3.3	x	[100]
Microtus arvalis	Common vole		3.0	x	[100]
Microtus guentheri	Vole		3.9	x	[143]
Microtus oeconomus	Root vole	1.4	1.8	x	[100]
Microtus pinetorum	Pine vole		2.8	x	[100]
Millardia gleadowi	Asian soft-furred rat		4.0	x	[229]
Millardia spp.	Asian soft-furred rat		4.0	x	[100]
Mus minutoides	Mouse		3.1	x	[17] [28] [191]
Mus musculus	House mouse	6.0		x	[21]
Mus musculus	Mouse		6.0	x	[143]
Myopus schisticolor	Wood lemming	1.0		x	[100]
Mystromys albicaudatus	White-tailed rat		6.0	x	[50]
Mystromys albicaudatus	White-tailed rat		6.0	x	[100]
Nannospalax spp.	Mediterranean blind mole-rat	4.5		x	[274]
Neotoma albigula	Wood rat		7.7	x	[143]
Neotoma floridana	Eastern woodrat	3.0		x	[100]
Notomys alexis	Australian hopping mouse		5.2	x	[143]
Notomys spp.	Australian hopping mouse		6.0	x	[100]
Nyctomys sumichrasti	Vesper rat		5.2	x	[143]
Ochrotomys nuttalli	Golden mouse		8.4	f	[229]
Ondatra zibethicus	Muskrat	3.0	10.0	x	[229]
Ondatra zibethicus	Muskrat	4.0		x	[100]
Onychomys leucogaster	Northern grasshopper mouse (+)		5.0	x	[100]
Onychomys torridus	Grasshopper mouse		4.6	x	[143]
Oxymycterus rutilans	Burrowing mouse		2.6	x	[143]
Pachyuromys duprasi	Fat-tailed gerbil		4.4	x	[100]
Pachyuromys duprasi	Fat-tailed gerbil		4.4	x	[143]
Peromyscus maniculatus	Deer mouse		8.3	x	[10]
Phloeomys spp.	Slender-tailed cloud rat		13.6	x	[276]
Phodopus sungorus	Striped hairy-footed hamster		3.2	x	[143]
Phodopus sungorus	Striped hairy-footed hamster (+)		3.0	x	[100]
Pithecheir spp.	Monkey-footed rat		3.0	x	[183]
Platacanthomys lasiurus	Spiny dormouse		1.7	f	[250]

	Pogonomys macrourus	Prehensile-tailed rat		2.4	x	[143]
	Praomys tullbergi	African soft-furred rat		5.2	x	[143]
	Pseudomys spp.	Australian native mouse		6.0	x	[100]
	Pseudomys australis	Australian native mouse		5.6	x	[143]
	Rattus rattus	Rat		4.2	x	[207]
	Reithrodon physodes	Coney rat		5.5	x	[143]
	Reithrodontomys spp.	American harvest mouse	1.5		x	[73]
	Rhabdomys pumilio	Four-striped grass mouse		2.9	x	[143]
	Rhizomys spp.	Bamboo rat		4.0	x	[176]
	Rhizomys spp.	Bamboo rat		4.0	x	[100]
	Rhombomys opimus	Great gerbil	4.0		f	[225] [263]
	Rhombomys opimus	Great gerbil	3.0		m	[225] [263]
	Saccostomus campestris	African pouched rat		2.8	x	[143]
	Saccostomus spp.	African pouched rat		3.0	x	[100]
	Sekeetamys calurus	Bushy-tailed jird		5.4	x	[100]
	Sekeetamys calurus	Bushy-tailed jird		5.4	x	[143]
	Sigmodon hispidus	Cotton rat		5.2	x	[143]
	Sundamys muelleri	Giant sunda rat (+)	0.5	2.0	x	[207]
	Synaptomys cooperi	Southern bog lemming (+)		2.5	x	[100]
	Synaptomys spp.	Bog lemming		2.4	f	[184]
	Tachyoryctes spp.	African mole-rat		3.1	x	[249]
	Tatera indica	Indian gerbil		7.0	x	[100]
	Tatera indica	Large naked-soled gerbil		7.0	x	[143]
	Thallomys paedulcus	Acacia rat		3.5	x	[143]
	Tylomys nudicaudus	Climbing rat		5.4	x	[143]
Myocastoridae	Myocastor coypus	Nutria		10.0	x	[186]
	Myocastor coypus	Nutria (+)	6.0		x	[21]
	Myocastor coypus	Nutria		6.0	x	[100]
Octodontidae	Octodon degus	Degus		7.1	x	[143]
	Octodontomys gliroides	Bori (+)		7.0	x	[100]
	Octodontomys gliroides	Chozchoz		7.3	x	[143]
	Spalacopus cyanus	Coruro		5.8	x	[143]
	Spalacopus cyanus	Coruro		6.0	x	[100]
Pedetidae	Pedetes capensis	Springhare		13.8	x	[143]
Sciuridae	Ammospermophilus leucurus	Antelope ground squirrel		5.8	x	[143]
	Atlantoxerus getulus	Barbary ground squirrel (+)		9.0	x	[100]
	Atlantoxerus getulus	Barbary ground squirrel		5.6	x	[143]
	Callosciurus erythraeus	Beautiful squirrel		16.1	x	[143]
	Callosciurus notatus	Plantain squirrel		9.6	x	[100]
	Cynomys ludovicianus	Prairie dog		8.5	x	[35] [100]
	Dremomys lokriah	Red-cheeked squirrel		7.1	x	[143]
	Eutamias (= Tamias) ruficaudus	Western American chipmunk	8.0		x	[11]
	Funisciurus congicus	African striped squirrel		9.5	x	[143]
	Glaucomys sabrinus	Northern flying squirrel		13.0	x	[100]
	Glaucomys volans	Southern flying squirrel (+)		12.0	x	[100] [143]
	Heliosciurus rufobrachium	Sun squirrel		8.9	x	[143]
	Marmota bobak	Bobak's marmot	15.0		x	[100]
	Marmota marmota	Alpine marmot		18.0	x	[100]
	Marmota marmota	Alpine marmot	15.0		x	[229]
	Menetes berdmorei	Berdmore's palm squirrel		4.3	x	[143]
	Petaurista petaurista	Giant flying squirrel		16.0	x	[100]

	Protoxerus stangeri	Oil palm squirrel		5.1	x	[143]
	Pteromys volans	Siberian flying squirrel	3.8		x	[21]
	Sciurus carolinensis	Eastern gray squirrel	23.5		x	[21]
	Sciurus carolinensis	Eastern gray squirrel (+)	23.6	12.0	x	[100]
	Sciurus carolinensis	Eastern gray squirrel	12.5	23.5	x	[143]
	Sciurus vulgaris	Eurasian red squirrel	12.0		x	[100]
	Spermophilus spp.	Ground squirrel (+)	11.0		f	[286]
	Tamias striatus	Eastern American chipmunk	3.0	8.0	x	[92] [100]
	Tamiasciurus spp.	Red squirrel	7.0	10.0	x	[100] [272] [10]
	Trogopterus xanthipes	Complex-toothed flying squirrel		12.0	x	[325]
	Xerus erythropus	Western ground squirrel		6.0	x	[100]
	Xerus inauris	African ground squirrel (+)		6.0	x	[229]
Spalacidae	Spalax spp.	Blind mole-rat (+)		4.0	x	[100]
Thryonomyidae	Thryonomys swinderianus	Cane rat	4.3		x	[21]
	Thryonomys swinderianus	Cane rat		4.3	x	[143]
Zapodidae	Napaeozapus insignis	Woodland jumping mouse	4.0		x	[229]
	Sicista betulina	Northern birch mouse	3.5		x	[100]
	Sicista spp.	Birch mouse	3.3		x	[130]
	Zapus hudsonius	Jumping mouse		5.0	x	[143]
Scandentia						
Tupaiidae	Ptilocercus lowii	Pen-tailed tree shrew		2.7	x	[176]
	Tupaia glis	Tree shrew		12.4	x	[143]
	Urogale everetti	Philippine tree shrew		11.5	x	[143]
Sirenia						
Dugongidae	Dugong dugon	Dugong (+)	55.0		x	[192]
	Dugong dugon	Dugong	70.0	10.0	x	[227] [128]
	Dugong dugon	Dugong	55.0		x	[100]
Trichechidae	Trichechus inunguis	Amazonian manatee		12.5	x	[127]
	Trichechus inunguis	Amazonian manatee (+)	30.0		x	[192]
	Trichechus manatus	Caribbean manatee	30.0		x	[21]
	Trichechus manatus	Caribbean manatee (+)		30.0	x	[143]
	Trichechus manatus	Caribbean manatee		28.0	x	[192]
Tubulidentata						
Orycteropodidae	Orycteropus afer	Aardvark	23.0		x	[21]
	Orycteropus afer	Aardvark		23.0	x	[100]
	Orycteropus afer	Aardvark		10.0	x	[192]
	Orycteropus afer	Aardvark		24.0	m	[289]
	Orycteropus afer	Aardvark		18.0	f	[289]
Xenarthra (= Edentata)						
Bradypodidae	Bradypus torquatus	Maned sloth	12.0		x	[100]
Dasypodidae	Calyptophraetus (= Chlamyphorus) retuses	Burmeister's armadillo	12.0		x	[21]
	Chaetophractus villosus	Larger hairy armadillo		19.0	x	[100]
	Chaetophractus villosus	Hairy armadillo (+)		20.0	x	[229]
	Dasypus novemcinctus	Common long-nosed armadillo	15.0		x	[100] [334]
	Euphractus sexcinctus	Six-banded armadillo		18.8	x	[143]
	Priodontes maximus	Giant armadillo (+)		4.0	x	[100]

	Priodontes maximus	Giant armadillo	15.0		x	[208]
	Tolypeutes spp.	Three-banded armadillo	15.0		x	[208]
	Zaedyus pichiy	Pichi		9.0	x	[143]
Megalonychidae	Choloepus didactylus	Linne's two-toed sloth (+)		27.8	x	[143]
	Choloepus hoffmanni	Hoffmann's two-toed sloth (+)		32.1	x	[143]
Myrmecophagidae	Cyclopes didactylus	Silky anteater		2.3	x	[208]
	Myrmecophaga tridactyla	Giant anteater		26.0	x	[100] [192]
	Myrmecophaga tridactyla	Giant anteater		25.8	x	[143]
	Tamandua tetradactyla	Southern tamandua (+)		9.0	x	[100]
	Tamandua mexicana	Northern tamandua		9.5	x	[192]
	Tamandua spp.	Tamandua		9.5	x	[143]

References

1. Ables, E.D., 1975. Ecology of the red fox in North America, in *Wild Canids*, M.W. Fox, Editor. New York: Van Nostrand Reinhold: p. 216-236.
2. Albignac, R., 1972. The carnivora of Madagascar, in *Biogeogaphy and Ecology in Madagascar*, R. Battistini and G. Richard-Vindard, Editors. The Hague: Dr. W. Junk B.B.: p. 667-682.
3. Albignac, R., 1975. Breeding the fossa *Cryptoprocta ferox* at Montpelier Zoo. *International Zoo Yearbook* 15: 147-150.
4. Alendal, E., 1984. Muskoxen in captivity in Europe and Asia, April 1983, in *Proceedings of First International Muskox Symposium*, D.R. Klein, R.G. White, and S. Keller, Editors. Biological Papers University Alaska Special Report: p. 9-11.
5. Allaro, M., V. Lebre, and J.-M. Robine, 1998. *Jean Calment, From Van Gogh's Time to Ours, 122 Extraordinary Years*. New York: W. H. Freeman and Company.
6. Ando, A., S. Shiraishi, and T.A. Uchida, 1987. Growth and development of the Smith's red-backed vole, *Eothenomys smithi. Journal Faculty Agriculture Kyushu University* 31: 309-20.
7. Archer, M., 1974. Regurgitation or mercyism in the western native cat, *Dasyurus geoffroii*, and the Red-tailed Wambenger, *Phascogale calura* (Marsupialia, Dasyuridae). *Journal of Mammology* 55: 448-52.
8. Asdell, S.A., 1964. *Patterns of Mammalian Reproduction*. Ithaca, NY: Cornell University Press.
9. Baker, R.C., F. Wilke, and C.H. Baltzo, 1970. The northern fur seal. *U.S. Department of Interior Circular* : 19.
10. Banfield, A.W.F., 1974. *The Mammals of Canada*: University Toronto Press.
11. Beg, M.A. and R.S. Hoffman, 1977. Age determination and variation in the red-tailed chipmunk, *Eutamias ruficaudus. Murrelet* 58: 26-36.
12. Bekoff, M., 1975. Social behavior and ecology of the African canidae: A review, in *The Wild Canids*, M.W. Fox, Editor. New York: Van Nostrand Reinhold: p. 120-142.
13. Bekoff, M., 1977. *Canis latrans. Mammalian Species* 79: 9.
14. Benjaminsen, T. and I. Christensen, 1979. The natural history of the bottlenose whale *Hyperodon ampullatus* (Forster), in *Behavior of Marine Animals*, H.E. Winn and B.L. Olla, Editors. New York: Plenum Press: p. 143-164.
15. Berger, J., 1986. *Wild Horses of the Great Basin*. Chicago: University of Chicago Press.
16. Bergurud, A.T., 1978. Caribou, in *Big Game of North America*, J.L. Schmidt and D.L. Gilbert, Editors. Harrisburg, PA: Stackpole Books: p. 83-101.
17. Berry, R.J., 1970. The natural history of the house mouse. *Field Studies* 3: 219-262.
18. Best, R.C. and H.D. Fisher, 1974. Seasonal breeding of the narwhal (*Monodon monoceros L.*). *Canadian Journal of Zoology* 52: 429-431.
19. Bigg, M.A., 1981. Harbour seal--*Phoca vitulina* and *P. largha*, in *Handbook of Marine Mammals*, S.H. Ridgway and R. Harrison, Editors. London: Academic Press: p. 1-27.

20. Blandford, P.R.S., 1987. Biology of the polecat *Mustela putorius*: A literature review. *Mammology Review* 17: 155-198.
21. Bobick, J.E. and M. Peffer, eds. 1993. *Science and Technology Desk Reference*. ed. Series. Bobick, J.E. and M. Peffer. Washington, D.C.: Gale Research Inc.
22. Bothma, J.D.P., 1971. Control and ecology of the black-backed jackal *Canis mesomelas* in the Transvaal. *Zoology Africa* 6: 187-193.
23. Bowen, W.D., D.J. Boness, and O.T. Oftedal, 1987. Mass transfer from mother to pup and subsequent mass loss by the weaned pup in the hooded seal, *Cystophora cristata*. *Canadian Journal of Zoology* 65: 1-8.
24. Bowles, D., 1986. Social behaviour and breeding of Babirusa, *Babyrousa babyrussa* at the Jersey Wildlife Preservation Trust. *Dodo* 23: 86-94.
25. Braham, H.W., 1984. Review of reproduction in the white whale, *Delphinapterus leucas* narwhal, *Monodon monocerus*, and Irrawaddy dolphin, *Orcaella brevirostres*, with comments on stock assessment, in *Reproduction in Whales, Dolphins, and Porpoises*, W.F. Perrin, J. R. L. Brownell, and D.P. DeMaster, Editors. Report International Whaling Commission: p. 81-89.
26. Breiwick, J.M., L.L. Eberhardt, and H.W. Braham, 1984. Population dynamics of western Arctic bowhead whales (*Balaena mysticetus*). *Canadian Journal of Fish and Aquatic Science* 41: 484-496.
27. Brodie, P.F., 1971. A reconsideration of aspects of growth, reproduction, and behavior of the white whale (*Delphinapterus leucas*), with reference to the Cumberland Sound, Baffin Island, population. *Journal of Fisheries Research Board Canada* 28: 1309-1318.
28. Bronson, F.H., 1979. The reproductive ecology of the house mouse. *Quarterly Review of Biology* 54: 265-299.
29. Brown, M.K., 1979. Two old beavers from the Adirondacks. *New York Fish and Game Journal* 26: 92.
30. Brownell, R.L.J., 1975. Progress report on the biology of the Franciscana dolphin, *Pontoporia blainvillei* in Uruguayan waters. *Journal of the Fisheries Reserve Board of Canada* 32: 1073-1078.
31. Bryden, M.M., R.J. Harrison, and R.J. Lear, 1977. Some aspects of the biology of *Peponocephala electra* (Cetacea: Delphinidae): 1. General and reproductive biology. *Australian Journal of Marine and Freshwater Research* 28: 703-715.
32. Bueler, L.E., 1973. *Wild Dogs of the World*. New York: Stein and Day.
33. Burns, J.J., 1981. Bearded seal--*Erignathus barbatus*, in *Handbook of Marine Mammals: The Walrus, Sea Lions, Fur Seals, and Sea Otter*, S.H. Ridgeway and R. Harrison, Editors. London: Academic Press: p. 145-170.
34. Burns, J.J., 1981. Ribbon seal--*Phoca fasciata*, in *Handbook of Marine Mammals: The Walrus, Sea Lions, Fur seals, and Sea Otter*, S.H. Ridgeway and R. Harrison, Editors. London: Academic Press: p. 89-109.
35. Burt, W.H. and R.P. Grossenheider, 1976. *A Field Guide to the Mammals*. Boston: Houghton Mifflin.

36. Caldwell, D.K., M.C. Caldwell, and D.W. Rice, 1966. Behavior of the sperm whale, *Physeter catodon* L., in *Whales, Dolphins, and Porpoises*, K.S. Norris, Editor. Berkeley: University of California: p. 678-717.
37. Carley, C.J., 1979. Status Summary: The Red Wolf (*Canis rufus*). Albuquerque: U. S. Fish and Wildlife Services.
38. Carpenter, C.R., 1965. The howlers of Barro Colorado Island, in *Primate Behavior: Field Studies of Monkeys and Apes*, I.D. Vore, Editor. New York: Holt, Rinehart and Winston: p. 250-291.
39. Caubere, B., P. Gaucher, and J.E. Julien, 1984. Un record mondial de longevite *in natura* pour un chiroptere insectivore? *Revue Ecologie* 39: 351-353.
40. Charles-Dominique, P., 1977. *Ecology and Behavior of Nocturnal Primates: Prosimians of Equatorial West Africa*. New York: Duckworth.
41. Chivers, D.J., 1977. The lesser apes, in *Primate Conservation*, P.M. Rainier III and G.H. Bourne, Editors. New York: Academic Press: p. 536-598.
42. Clark, C.W., 1985. Economic aspects of marine mammal fishery interactions, in *Marine Mammals and Fisheries*, J.R. Beddington, R.J.H. Beverton, and D.M. Lavigne, Editors. London: George Allen and Unwin Publishers: p. 34-38.
43. Collett, S.F., 1981. *Population Characteristics of Agouti paca (Rodentia) in Columbia* Vol. 5: Michigan State University Museum Publications Biology Series.
44. Collins, L.R., 1973. *Monotremes and Marsupials*. Washington, D. C.: Smithsonian Institute Press.
45. Corbet, G.B. and H.N. Southern, 1977. *The Handbook of British Mammals*. London: Blackwell Scientific Publications.
46. Craighead, J.J., J. F. C. Craighead, and J. Sumner, 1976. Reproductive cycles and rates in the grizzly bear, *Ursus arctos horribilis*, of the Yellowstone ecosytem, in *Bears--Their Biology and Management*, M.R. Pelton, J.W. Lentfer, and G.E. Folk, Editors. International Union Conservation Nature Publ.: p. 337-356.
47. Currier, M.J.P., 1983. *Felis concolor. Mammalian Species* 200: 7.
48. Cuttle, P., 1982. Life history of the dasyurid marsupial *Phascogale tapoatafa* in *Carnivorous Marsupials*, M. Archer, Editor. Mosman: Roy. Zool. Soc. New South Wales: p. 13-22.
49. Dagg, A.I. and J.B. Foster, 1976. *The Giraffe: Its Biology, Behavior, and Ecology*. New York: Van Nostrand Reinhold.
50. Dean, W.R.J., 1978. Conservation of the white-tailed rat in South Africa. *Biological Conservation* 13: 133-140.
51. DeMaster, D.P. and I. Stirling, 1981. *Ursus maritimus. Mammalian Species* 145: 7.
52. Dittrich, L., 1967. Breeding Kirk's dik-dik *Madoqua kirki thomasi* at Hanover Zoo. *International Zoo Yearbook* 7: 171-173.
53. Dittrich, L., 1972. Gestation periods and age of sexual maturity of some African antelopes. *International Zoo Yearbook* 12: 171-173.
54. Doherty, J., 1989. Twin babyrusas born at the New York Zoological Park. *American Association of Zoological Parks Aquarium Newsletter* 30: 21.

55. Doyle, G.A. and S.K. Bearder, 1977. The galagines of South Africa, in *Primate Conservation*, P.M. Rainier III and G.H. Bourne, Editors. New York: Academic Press: p. 1-35.

56. Earle, R.D. and K.R. Kramm, 1980. Techniques for age determination in the Canadian porcupine. *Journal of Wildlife Management* 44: 413-419.

57. Eaton, R.L., 1974. *The Cheetah*. New York: Van Nostrand Reinhold.

58. Egbert, A.L. and A.W. Stokes, 1976. The social behavior of brown bears on an Alaskan salmon stream, in *Bears--Their Biology and Management*, M.R. Pelton, J.W. Lentfer, and G.E. Folk, Editors. International Union Conservation Nature Publ.: p. 41-56.

59. Egoscue, H.J., 1979. *Vulpes velox. Mammalian Species* 122: 5.

60. Egoscue, H.J., J.G. Bittmenn, and J.A. Petrovich, 1970. Some fecundity and longevity records for captive small mammals. *Journal of Mammology* 51: 622-623.

61. Eisenberg, J.F., 1975. Tenrecs and solenodons in captivity. *International Zoo Handbook* 15: 6-12.

62. Eisenberg, J.F., 1981. *The Mammalian Radiations: An Analysis of Trends in Evolution, Adaptation, and Behavior*: The University of Chicago Press.

63. Eisenberg, J.F. and E. Gould, 1970. Behavior of *Solenodon paradoxus* in captivity with comments on the behavior of other insectivora. *Zoologica* 51: 49-58.

64. Eisenberg, J.F. and E. Maliniak, 1974. The reproduction of the genus microgale in captivity. *International Zoo Yearbook* 14: 108-110.

65. Ellis, R., 1980. *The Book of Whales* 2nd ed. New York.

66. Everard, C.O.R. and E.S. Tikasingh, 1973. Ecology of the rodents, *Proechimys guyannensis trinitatis* and *Oryzomys capito velutinus*, on Trinidad. *Journal of Mammalogy* 54: 875-876.

67. Ewer, R.F., 1973. *The Carnivores*. Ithaca, NY: Cornell University Press.

68. Fay, F.H., 1981. Walrus *Odobenus rosmarus*, in *Handbook of Marine Mammals: The Walrus, Sea Lions, Fur Seals, and Sea Otter*, S.H. Ridgeway and R. Harrison, Editors. London: Academic Press: p. 1-23.

69. Fay, F.H., 1982. Ecology and biology of the Pacific walrus, *Odobenus rosmarus divergens. North American Fauna* : 279.

70. Fay, F.H., 1985. *Odobenus rosmarus. Mammalian Species* 238: 7.

71. Fedigan, L., 1991. Life span and reproduction in Japanese macaquefemales, in *The Monkeys of Arashiyama: 35 Years of Research in Japan and the West*, L.M. Fedigan and P.J. Asquith, Editors. Albany, NY: University of New York Press.

72. Feldhamer, G.A., K.C. Farris-Renner, and C.M. Barker, 1988. *Dama dama. Mammalian Species* 128: 7.

73. Fisler, G.F., 1971. Age structure and sex ratio in populations of *Reithrodontomys. Journal of Mammalogy* 52: 653-662.

74. Fleming, T.H., 1971. Population Ecology of Three Species of Neotropical Rodents *Miscellaneous Publications Museum Zoology University of Michigan:* University of Michigan.

75. Fleming, T.H., 1974. The population ecology of two species of Costa Rican Heteromyid rodents. *Ecology* 55: 493-510.

76. Fooden, J., 1979. Taxonomy and evolution of the *Sinica* group of macaques: I. Species and sub-species accounts of *Macaca sinica. Primates* 10: 109-140.
77. Frame, L.H., J.R. Malcolm, G.W. Frame, and H.V. Lawick, 1979. Social organization of African wild dogs (*Lycaon pictus* on the Serengeti plains, Tanzania (1967-1978). *Zeitschrift Tierpsychologia.* 50: 225-249.
78. Franklin, W.L., 1974. The social behavior of the vicuna, in *The Behavior of Ungulates and its Relation to Management*, V. Geist and F. Walther, Editors. International Union Conservation Nature Publ.: p. 477-487.
79. Franklin, W.L., 1982. Biology, ecology, and relationship to man of the South American camelids. *Pymatuning Laboratory Ecology Special Publication* 6: 457-489.
80. Frazer, J.F.D., 1973. Specific foetal growth rates of cetaceans. *Journal of Zoology* 169: 111-126.
81. Froehlich, J.W., J. R.W. Thorington, and J.S. Otis, 1981. The demography of howler monkeys (*Alouatta palliata*) on Barro Colorado Island, Panama. *International Journal of Primatology* 2: 208-236.
82. Frost, K.J. and L.F. Lowry, 1981. Ringed, baikal and caspian seals-*Phoca hispida, Phoca sibirica* and *Phoca Caspica*, in *Handbook of Marine Mammals: The WalrusSea Lions, Fur Seals and Sea Otter*, S.H. Ridgeway and R. Harrison, Editors. London: Academic Press: p. 29-53.
83. Gaisler, J. and V. Hanak, 1969. Summary of the results of bat banding in Czechoslovakia--1948-1967. *LynxPraha* 10: 25-34.
84. Gaskin, D.E., 1977. Harbour porpoise *Phocena phocena* (L.) in the western approaches to the Bay of Fundy 1969-1975. *Report of the International Whaling Commission* 27: 487-492.
85. Gaskin, D.E. and B.A. Blair, 1977. Age determination of harbour porpoise, *Phocena phocena* (L.) in the western North Atlantic. *Canadian Journal of Zoology* 55: 18-30.
86. Gaskin, D.E., G.J.D. Smith, A.P. Watson, W.Y. Yasui, and D.B. Yurick, eds. 1984. *Reproduction in the Porpoises (Phocenidae): Implications for Management*. Reproduction in Whales, Dolphins and Porpoises ed. Series. Gaskin, D.E., G.J.D. Smith, A.P. Watson, W.Y. Yasui, and D.B. Yurick. Vol. Special Issue No. 6 Report of the International Whaling Commission.
87. Gauthier-Pilters, H. and A.I. Dagg, 1981. *The Camel.* University of Chicago Press.
88. Geist, V., 1971. *Mountain Sheep: A Study in Behavior and Evolution*: University of Chicago Press.
89. George, W., 1978. Reproduction in female gundis (Rodentia: Ctenodactylidae). *Journal of Zoology* 185: 57-71.
90. Gier, H.T., 1975. Ecology and behavior of the coyote (*Canis latrans*), in *The Wild Canids: Their Systematics, Behavioral Ecology and Evolution*, M.W. Fox, Editor. New York: Van Nostrad Reinhold Company: p. 247-262.
91. Glenn, L.P., J.W. Lentfer, J.B. Faro, and L.H. Miller, 1976. Reproductive biology of female brown bears, *Ursus arctos*, McNeil River, Alaska, in *Bears--Their Biology and Management*, M.R. Pelton, J.W. Lentfer, and G.E. Folk, Editors. International Union Conservation Nature Publ.: p. 381-390.

92. Godin, A.J., 1977. *Wild Mammals of New England.* Baltimore: John Hopkins University Press.

93. Goldman, C.A., 1987. *Crossarchus obscurus. Mammalian Species* 290: 5.

94. Goodall, J., 1988. *In the Shadow Of Man.* Boston: Houghton Mifflin Co.

95. Green, M.J.B., 1987. Scent-marking in the Himalayan musk deer (*Moschus chrysogaster*). *Journal of Zoology* 1: 721-737.

96. Griffiths, M., 1978. *The Biology of Monotremes.* New York: Academic Press.

97. Griner, L.A., 1980. East Bornean proboscis monkey, *Nasalis larvatus orientalis. Zool. Garten N. F. Jena* 50: 73-81.

98. Groves, C.P., 1972. *Ceratotherium simum. Mammalian Species* 8: 8.

99. Grubb, P., 1981. *Equus burchelli. Mammalian Species.* 157: 9.

100. Grzimek, B., ed. 1990. *Grzimek's Animal Life Encyclopedia. Mammals I - IV.* ed. Series. Grzimek, B. Vol. I-IV. New York: McGraw-Hill Publishing Company.

101. Grzimek, N.C.B., ed. 1975. Grzimek's Animal Life Encyclopedia ed. Series. Grzimek, N.C.B. Vol. 1-5, Mammals. New York, N.Y.: Van Nostrand Reinhold Co.

102. Guggisberg, C.A.W., 1975. *Wild Cats of the World.* New York: Taplinger.

103. Guiler, E.R. and D.A. Kitchener, 1967. Further observations on longevity in the Wild Potoroo *Potorous tridactylus. Australian Journal of Science* 30: 105-106.

104. Hakeem, A., R. Sandoval, M. Jones, and J. Allman, 1996. Brain and lIfe span in primates, in *Handbook of the Psychology of Aging.*, J. Birren, Editor. Academic Press: p. 78-104.

105. Haley, D., 1978. *Marine Mammals of Eastern North Pacific and Arctic Waters.* Seattle: Pacific Research Press.

106. Harrison, D.L., 1972. *The Mammals of Arabia.* London: Ernest Benn.

107. Harrison, R.J., M.M. Bryden, and D.A. McBrearty, 1981. The ovaries and reproduction in *Pontoporia blainvillei* (Cetacea: Platanistidae). *Journal of Zoology* 193: 563-580.

108. Harrison, R.J. and J. R. L. Brownell, 1971. The gonads of the South American dolphins, *Inia geoffrensis, Pontoporia blainvillei* and *Sotalia fluviatilis. Journal of Mammalogy* 52: 413-419.

109. Harvey, M.J., 1976. Home range, movements, and diet activity of the eastern mole, *Scalopus aquaticus. American Midland Naturalist* 95: 436-445.

110. Harvey, P.H., R.D. Martin, and T.H. Clutton-Brock, 1987. Life histories in comparative perspective, in *Primate Societies*, B.B. Smuts, D.L. Cheney, R.M. Seyfarth, R.W. Wrangham, and T.T. Struhsaker, Editors. Chicago: The University of Chicago Press: p. 181-196.

111. Hasler, J.F., 1975. A review of reproduction and sexual maturation in the microtine rodents. *Biologist* 57: 52-86.

112. Haugen, A.O., 1974. Reproduction in the plains buffalo. *Iowa State Journal of Research* 49: 1-8.

113. Hayden, P. and R.G. Lindberg, 1976. Survival of laboratory-reared pocket mice, *Perognathus longimembris. Journal of Mammalogy* 57: 266-272.

114. Hendrichs, H., 1975. Changes in a population of dik-dik, *Madoqua (Rhynchotragus) kirki* (Gunther 1880). *Zeitschrift Tierpsychologia.* 38: 55-69.

115. Hill, C., 1975. The longevity record for *Colobus*. *Primates* 16: 235.

116. Hill, W.C.O., 1970. *Comparative Anatomy and Taxonomy VIII: Cynopithecinae, Papio, Mandrillus, Theropithecus*. New York: Wiley-Interscience.

117. Hillman, C.N. and J.W. Carpenter, 1980. Masked mustelid. *Nature Conservancy News* 30: 20-23.

118. Hindell, M.A. and G.J. Little, 1988. Longevity, fertility, and philopatry of two male southern elephant seals (*Mirounga leonina*) at Macquarie Island. *Marine Mammal Science* 4: 168-171.

119. Hitchcock, H.B., 1965. Twenty-three years of bat banding in Ontario and Quebec. *Canadian Field Naturalist* 79: 4-14.

120. Hoeck, H.N., 1975. Differential feeding behaviour of the sympatric hyrax *Procavia johnstoni* and *Heterohyrax brucei*. *Oecologia* 22: 15-47.

121. Hoeck, H.N., 1982. Population dynamics, dispersal and genetic isolation in two species of hyrax (*Heterohyrax brucei* and *Procavia johnstoni*) on habitat islands of the Serengeti. *Zeitschrift Tierpsychologia.* 59: 177-210.

122. Hoeck, H.N., 1984. Hyraxes, in *The Encyclopedia of Mammals*, D. MacDonald, Editor. New York: Facts on File Publ.: p. 462-465.

123. Hoeck, H.N., 1989. Demography and competition in hyrax. *Oecologia* 79: 353-360.

124. Hoeck, H.N., H. Klein, and P. Hoeck, 1982. Flexible social organization in hyrax. *Zeitschrift Tierpsychologia.* 59: 265-298.

125. Horwich, R.H., 1989. Cyclic development of contact behavior in apes and humans. *Primates* 30: 269-279.

126. Hunsaker, D.H., 1977. Ecology of new world marsupials, in *The Biology of Marsupials*, D.H. Hunsaker, Editor. New York: Academic Press: p. 95-156.

127. Husar, S.L., 1977. *Trichechus inunguis. Mammalian Species* 72: 4.

128. Husar, S.L., 1978. *Dugong dugon. Mammalian Species* 88: 7.

129. Husson, A.M., 1978. *The Mammals of Suriname*. Leiden: E. J. Brill.

130. Ivanter, E.V., 1972. Contribution to the ecology of *Sicista betulina* pall. *Aquilo, Series Zoology* 13: 103-108.

131. Izard, M.K., 1987. Lactation length in three species of *Galago*. *American Journal of Primatology* 13: 73-76.

132. Jackson, H.H.T., 1961. *Mammals of Wisconsin*. Madison: University Wisconsin Press.

133. Jackson, R. and G.G. Ahlborn, 1988. *Observations on the ecology of snow leopard in West Nepal*. in *Proceedings 5th International Snow Leopard Symposium*.

134. Jarman, P.J. and M.V. Jarman, 1973. Social behaviour, population strucure and reproductive potential in impala. *East African Wildlife Journal* 11: 329-338.

135. Jarvis, J.U.M., 1969. The breeding season and litter size of African mole-rats. *Journal of Reproduction and Fertility Supplement* 6: 237-248.

136. Jarvis, J.U.M., 1973. Activity patterns in the mole-rats *Tachyoryctes splendens* and *Heliophobius argenteocinereus*. *Zoology Africa* 8: 101-119.

137. Jarvis, J.U.M., 1978. Energetics of survival in *Heterocephalus glaber* (Ruppel), the naked mole-rat (Rodentia Bathyergidae). *Bulletin of the Carnegie Museum of Natural History* : 81-87.
138. Jarvis, J.U.M., 1985. Ecological studies on *Heterocephalus glaber*, the naked mole-rat, in Kenya. *National Geographic Society Research Report* 20: 429-337.
139. Jarvis, J.U.M. and B. Sale, 1971. Burrowing and burrow patterns of east African mole-rats *Tachyoryctes, Heliophobius* and *Heterocephalus. Journal of Zoology* 163: 451-479.
140. Jefferson, T.A., 1988. *Phocoenoide dalli. Mammalian Species* 319: 7.
141. Jones, M.L., 1968. Longevity of primates in captivity. *International Zoo Yearbook* 8: 183.
142. Jones, M.L., 1972. Unpublished: . Hotel Saint Mark, San Francisco.
143. Jones, M.L., 1982. Longevity of captive mammals. *Zoological Garten N. F. Jena* 52: 113-28.
144. Jones, M.L. and V.J.A. Manton, 1983. History in captivity, in *The Biology and Management of an Extinct Species Pere David's Deer.* B.B. Beck and C. Wemmer, Editors. Park Ridge, NJ: Noyes Publ.: p. 1-14.
145. Jonkel, C.J., 1978. Black, brown (grizzly) and polar bears, in *Big Game of North America*, J.L. Schmidt and D.L. Gilbert, Editors. Harrisburg: Stackpole Books: p. 227-248.
146. Kasuya, T., 1978. The life history of Dall's porpoise with special reference to the stock of the Pacific coast of Japan. *Scientific Report Whales Research Institute* 30: 1-63.
147. Kasuya, T. and Aminul-Haque, K. M., 1972. Some information on distribution and seasonal movement of the Ganges dolphin. *Sci. Report Whales Res. Inst.* 24: 109-115.
148. Kasuya, T. and K. Kureha, 1979. The population of finless porpoise in the inland Sea of Japan. *Scientific Report Whales Research Institute* 31: 1-44.
149. Kasuya, T. and H. Marsh, 1984. Life history and reproductive biology of the short-finned pilot whale, *Globicephala macrorhynchus,* off the Pacific coast of Japan, in *Reproduction in Whales, Dolphins, and Porpoises*, W.F. Perrin, R.L.J. Brownell, and D.P. DeMaster, Editors. Report of the International Whaling Commission: p. 259-310.
150. Kasuya, T. and J. R. L. Brownell, 1979. Age determination, reproduction, and growth of Franciscana dolphin, *Pontoporia blainvillei. Science Report Whales Research Institute* 31: 45-67.
151. Keen, R. and H.B. Hitchcock, 1980. Survival and longevity of the little brown bat (*Myotis lucifugus*) in southeastern Ontario. *Journal of Mammalogy* 61: 1-7.
152. Kellas, L.M., 1955. Observations on the reproductive activities, measurements, and growth rate of the dik-dik (*Rhynchotragus kirkii thomasi* Neumann). *Proceedings of the Zoological Society of London* 124: 751-784.
153. Kennelly, J.J., 1978. Coyote reproduction, in *Coyotes: Biology, Behavior, and Management*, M. Bekoff, Editor. New York: Academic Press: p. 73-93.
154. Kenyon, K.W., 1981. Monk seals--*Monachus*, in *Handbook of Marine Mammals: The Walrus, Sea Lions, Fur Seals, and Sea Otter*, S.H. Ridgeway and R. Harrison, Editors. London: Academic Press: p. 195-220.
155. Kerr, M.A., 1965. The age at sexual maturity in male impala. *Arnoldia* 1: 1-6.

156. Kilgore, D.L.J., 1969. An ecological study of the swift fox (*Vulpes velox*) in the Oklahoma panhandle. *American Midland Naturalist* 81: 512-534.

157. King, J.E., 1983. *Seals of the World*. London: British Museum Natural History.

158. Kingdon, J., 1974. *An Atlas of Evolution in Africa II(B): Hares and Rodents* Vol. II(B). Hares and Rodents. London: Academic Press.

159. Kingdon, J., 1977. *An Atlas of Evolution in Africa III (A): Carnivores*. London: Academic Press.

160. Kingdon, J., 1979. *An Atlas of Evolution in Africa III(B): Large Mammals*. London: Academic Press.

161. Kingdon, J., 1982. *An Atlas of Evolution in Africa III (C and D)* Vol. III (C and D). London: Academic Press.

162. Kirkpatrick, T.H., 1965. Studies of Macropodidae in Queensland: 2. Age estimation in the grey kangaroo, the eastern wallaroo and the red-necked wallaby, with notes on dental abnormalities. *Queensland Journal of Agricultural Animal Science* 22: 301-317.

163. Klingel, H., 1974. A comparison of the social behaviour of the Equidae, in *The Behaviour of Ungulates and It's Relation to Management*, V. Geist and F. Walther, Editors. International Union Conservation Nature Publ.: p. 124-132.

164. Kohncke, M. and K. Leonhardt, 1986. *Cryptoprocta fossa. Mammalian Species* 254: 1-5.

165. Kooyman, G.L., 1981. Crabeater seal *Lobodon carcinophagus*, in *Handbook of Marine Mammals*, S.H. Ridgway and R. Harrison, Editors. London: Academic Press: p. 221-235.

166. Kooyman, G.L., 1981. Leopard seal--*Hydrurga leptonyx*, in *Handbook of Marine Mammals*, S.H. Ridgway and R. Harrison, Editors. London: Academic Press: p. 261-274.

167. Krasinski, Z.A., 1978. Dynamics and structure of the European bison population in the Bialowieza primeval forest. *Acta Theriologica* 23: 3-48.

168. Lacher, T.E.J., 1981. The comparative social behavior of *Kerodon rupestris* and *Galea spixii* and the evolution of behavior in the Caviidae. *Bulletin of the Carnegie Museum of Natural History* : 71.

169. Latinen, K., 1987. Longevity and fertility of the polar bear, *Ursus maritimus* Phipps, in captivity. *Zool. Garten* 57: 197-199.

170. Laurie, A., 1982. Behavioural ecology of the greater one-horned rhinoceros (*Rhinoceros unicornis*). *Journal of Zoology* 196: 307-341.

171. Laurie, A. and J. Seidensticker, 1977. Behavioural ecology of the sloth bear (*Melursus ursinus*). *Journal of Zoology* 182: 187-204.

172. Laurie, W.A., E.M. Lang, and C.P. Groves, 1983. *Rhinoceros unicornis. Mammalian Species* 211: 6.

173. Laursen, L. and M. Bekoff, 1978. *Loxodonta africana. Mammalian Species* 92: 8.

174. LaVal, R.K., 1973. Observations on the biology of *Tadarida brasilienses cynocephala* in southeastern Louisiana. *American Midland Naturalist* 89: 112-120.

175. Leatherwood, S. and R.R. Reeves, 1978. Porpoises and dolphins, in *Marine Mammals of Eastern North Pacific and Arctic Waters*, D. Haley, Editor. Seattle: Pacific Search Press: p. 97-111.

176. Lekagul, B. and J.A. McNeely, 1977. *Mammals of Thailand*. Bangkok: Sahakarnbhat.

177. Leuthold, W., 1974. Observations on home range and social organization of lesser kudu, *Tragelaphus imberbis* (Blyth, 1869), in *The Behavior of Ungulates and Its Relation to Management*, V. Geist and F. Walther, Editors. International Union Conservation Nature Publ.: p. 206-234.

178. Leuthold, W., 1978. On social organization and behaviour of the gerenuk *Litocranius walleri* (Brooke 1878). *Zeitschrift Tierpsychologia* 47: 194-216.

179. Leuthold, W., 1978. On the ecology of the gerenuk *Litocranius walleri*. *Journal of Animal Ecology* 47: 561-580.

180. Leuthold, W., 1979. The lesser kudu, *Tragelaphus imberbis* (Blyth, 1869): The ecology and behaviour of an African antelope. *Saugertierk Mitt.* 27: 1-75.

181. Lewis, J.G. and R.T. WIlson, 1979. The ecology of Swayne's hartebeest. *Biological Conservation* 15: 1-12.

182. Lewis, S.E., 1994. Night roosting ecology of pallid bats (*Antrozous pallidus*) in Oregon. *American Midland Naturalist* 132: 219-226.

183. Lim, B.L. and I. Muul, 1975. Notes on a rare species of arboreal rat, *Pithecheir parvus* Kloss. *Malayan Nature* 28: 181-185.

184. Linzey, A.V., 1983. *Synaptomys cooperi*. *Mammalian Species* 210: 5.

185. Lockyer, C., 1984. Review of baleen whale (Mysticeti) reproduction and implications for management, in *Reproduction in Whales, Dolphins, and Porpoises*, W.F. Perrin, J. R. L. Brownell, and D.P.D. Master, Editors. Report International Whaling Commission: p. 27-50.

186. Lockyer, C., R.N.P. Goodall, and A.R. Galeazzi, 1988. Age and body-length characteristics of *Cephalorhynchus commersonii* from incidentally caught specimens off Tierra Del Fuego, in *Report of the International Whaling Commission: Biology of the Genus Cephalorhynchus*, J. R.L. Brownell and G.P. Donovan, Editors. Cambridge International Whaling Commission: p. 344.

187. Lopez-Forment, W., 1980. Longevity of wild *Desmodus rotundus* in Mexico. in *Proceedings 5th International Bat Research Conference*.

188. Lott, D.F., 1981. Sexual behavior and intersexual strategies in American bison. *Zeitschrift Tierpsychologia* 56: 97-114.

189. Loughlin, T.R., M.A. Perez, and R.L. Merrick, 1987. *Eumetopias jubatus*. *Mammalian Species* 283: 7.

190. Lovejoy, B.P. and H.C. Black, 1979. Population analysis of the mountain beaver, *Aplodontia rufapacifica*, in western Oregon. *Northwest Science* 53: 82-89.

191. Lowery, G.H.J., 1974. *The Mammals of Louisiana and Its Adjacent Waters*: Louisiana State University Press.

192. MacDonald, D., ed. 1984. *The Encyclopedia of Mammals*. ed. Series. MacDonald, D. New York: Facts on File, Inc.

193. Malcolm, J.R., 1980. African wild dogs play every game by their own rules. *Smithsonian* 11: 62-71.

194. Mansfield, A.W., T.G. Smith, and B. Beck, 1975. The narwhal, *Monodon monoceros*, in eastern Canadian waters. *Journal of Fisheries Research Board Canada* 32: 1041-1046.

195. Marquette, W.M., 1978. Bowhead whale in *Marine Mammals of Eastern North Pacific and Arctic Waters*, D. Haley, Editor. Seattle, WA: Pacific Search Press: p. 70-81.

196. Maynes, G.M., 1977. Distribution and aspects of the biology of the Parma Wallaby, *Macropus parma*, in New South Wales. *Australian Wildlife Research* 4: 109-125.

197. Mazak, V., 1981. *Panthera tigris. Mammalian Species* 152: 8.

198. McLaren, I.A., 1984. True seals, in *The Encyclopedia of Mammals*, D. Macdonald, Editor. New York: Facts on File Publ.: p. 270-79.

199. Mead, J.G., 1984. Survey of reproductive data for the beaked whales (Ziphiidae), in *Reproduction in Whales, Dolphins, and Porpoises*, W.F. Perrin, J. R. L. Brownell, and D.P. DeMaster, Editors. Report International Whaling Commission: p. 91-96.

200. Meagher, M.M., 1973. The Bison of Yellowstone National Park: U.S. National Park Service.

201. Mech, L.D., 1970. *The Wolf: The Ecology and Behavior of an Endangered Species*. Minneapolis: University of Minnesota Press.

202. Mech, L.D., 1974. *Canis lupus. Mammalian Species* 37: 6.

203. Mech, L.D., 1977. Population trend and winter deer consumption in a Minnesota wolf pack, in *Proceedings of the 1975 Predator Symposium*, R.L. Phillips and C. Jonkel, Editors. Univ. Montana, Missoula: Montana Forest and Conservation Experimental Station: p. 53-83.

204. Mech, L.D., 1977. Productivity, mortality, and population trends of wolves in northeastern Minnesota. *Journal of Mammalogy* 58: 559-574.

205. Mech, L.D., 1989. Wolf longevity in the wild. *Endangered Species Technical Bulletin* 14: 8.

206. Medjo, D.C. and L.D. Mech, 1976. Reproductive activity in nine- and ten-month-old wolves. *Journal of Mammalogy* 57: 406-408.

207. Medway, L., 1978. *The Wild Mammals of Malaya (Peninsular Malaysia) and Singapore*. Kuala Lumpur: Oxford University Press.

208. Merret, P.K., 1983. Edentates. *Zoological Trust of Guernsey* : 6.

209. Michalak, I., 1983. Reproduction, maternal and social behaviour of the European water shrew under laboratory water conditions. *Acta Theriologica* 28: 3-24.

210. Millar, J.S. and F.C. Zwickel, 1972. Determination of age, age structure, and mortality of the pika, *Ochotona princeps* (Richardson). *Canadian Journal of Zoology* 50: 229-232.

211. Mills, M.G.L., 1982. *Hyaena brunnea. Mammalian Species* 194: 5.

212. Mitchell, E.D., 1975. *Report of the meeting on smaller cetaceans*. in *Meeting on Smaller Cetaceans*. Montreal, Canada: Journal Fisheries Research Board Canada.

213. Mitchell, E.D., 1978. Finn whales, in *Marine Mammals of Eastern North Pacific and Arctic Waters*, D. Haley, Editor. Seattle: Pacific Search Press: p. 36-45.

214. Mock, O.B. and C.H. Connaway, 1975. Reproduction of the least shrew (*Cryptotis parva)* in captivity, in *The Laboratory Animal in the Study of Reproduction*, T. Antikatzides, S. Erichsen, and A. Spiegel, Editors. Stuttgart: Fischer Verlag: p. 59-74.

215. Moehlman, P.D., 1978. Jackals of the Serengeti. *Wildlife News; African Wildlife Leadership Foundation* 13: 2-6.

216. Moehlman, P.D., 1983. Socioecology of silverbacked and golden jackals (*Canis mesomelas* and *Canis aureus*). *American Society of Mammals Special Publication* 7: 423-453.

217. Moehlman, P.D., 1987. Social organization in jackals. *American Science* 75: 366-375.

218. Mondolfi, E. and R. Hoogesteijn, 1986. Notes on the biology and status of the jaguar in Venezuela, in *Cats of the World: Biology, Conservation, and Management*, S.D. Miller and D.D. Everett, Editors. Washington, D. C.: National Wildlife Federation: p. 85-123.

219. Morton, S.R., 1978. An ecological study of *Sminthopsis crassicaudata* (Marsupialia: Dasyuridae). III. Reproduction and life history. *Australian Wildlife Reseach* 5: 183-211.

220. Moss, C.J., 1983. Oestrous behaviour and female choice in the African elephant. *Behaviour* 86: 167-196.

221. Murie, A., 1981. The grizzlies of Mount McKinley. *U.S. National Park Service Science Monograph* 14: 251.

222. Musser, G.G., 1979. Results of the Archbold expeditions. No. 102. The species of *Chiropodomys,* aboreal mice of Indochina and the Malay Archipelago. *Bulletin of the American Museums of Natural History* 162: 377-445.

223. Napier, J.R. and P.H. Napier, 1967. *A Handbook of Living Primates*. New York: Academic Press.

224. Nash, L.T. and C.S. Harcourt, 1986. Social organization of galagos in Kenyan coastal forests: II. *Galago garnettii. American Journal of Primatology* 10: 357-69.

225. Naumov, N.P. and V.S. Lobachev, 1975. Ecology of desert rodents of the U.S.S.R., in *Rodents in Desert Environments*, I. Prakash and P.K. Ghosh, Editors. The Hague: W. Junk: p. 465-598.

226. Nerini, M.K., H.W. Braham, W.M. Marquette, and D.J. Rugh, 1984. Life history of the bowhead whale, *Balaena mysticetus* (Mammalia: Cetacea). *Journal of Zoology* 204: 443-468.

227. Nishiwaki, M. and H. Marsh, 1985. Dugong--*Dugong dugon* in *Handbook of Marine Mammals*, S.H. Ridgway and R. Harrison, Editors. London: Academic Press: p. 1-31.

228. Novikov, G.A., 1962. *Carnivorous Mammals of the U. S. S. R.* Jerusalem: Israel Progr. Sci. Transl.

229. Nowak, R.M., 1991. *Walker's Mammals of the World, Vol. I* 5th ed Vol. I. Baltimore: Johns Hopkins University Press.

230. Odell, D.K., 1975. Status and aspects of the life history of the bottlenose dolphin, *Tursiops truncatus* in Florida. *Journal of Fisheries Research Board Canada* 32: 1055-1058.

231. Ohsumi, S., 1966. Sexual segregation of the sperm whale in the North Pacific. *Scientific Report Whales Research Institute* 20: 1-16.

232. Ohsumi, S., T. Kasuya, and M. Nishiwaki, 1963. The accumulation rate of dentinal growth layers in the maxillary tooth of the sperm whale. *Scientific Report Whales Research Institute, Tokyo* 17: 15-36.

233. Ohtaishi, N. and M. Yoneda, 1981. A Thirty Four Years Old Male Kuril Seal From Shiretoko Peninsula. *Hokkaido Scientific Report Whales Research Institute* 33: 131-135.

234. Owen-Smith, N., 1974. The social system of the white rhinoceros, in *The Behavior of Ungulates and Its Relation to Management*, V. Geist and F. Walther, Editors. International Union Conservation Nature Publ.: p. 341-351.

235. Owen-Smith, N., 1984. Rhinoceros, in *The Encyclopedia of Mammals*, D. Macdonald, Editor. New York: Facts on File Publ.: p. 490-497.

236. Parish, A., 1999. Personal communication to Debra S. Judge, Dept. Anthropology, University California at Davis.

237. Pearson, A.M., 1976. Population characteristics of the Arctic mountain grizzly bear, in *Bears--Their Biology and Management*, M.R. Pelton, J.W. Lentfer, and G.E. Folk, Editors. International Union Conservation Nature Publ.: p. 247-260.

238. Pearson, O.P., S. Martin, and J. Bellati, 1987. Demography and reproduction of the silky desert mouse (*Eligmodontia*) in Argentina. *Fieldiana Zoologia* 39: 443-446.

239. Penzhorn, B.L., 1985. Reproductive characteristics of a free-ranging population of Cape Mountain zebra (*Equus zebra zebra*). *Journal of Reproduction and Fertility* 73: 51-57.

240. Perrin, W.F. and S.B. Reilly, 1984. Reproductive parameters of dolphins and small whales of the family Delphinidae, in *Reproduction in Whales, Dolphins, and Porpoises*, W.F. Perrin, J. R. L. Brownell, and D.P. DeMaster, Editors. Report International Whaling Commission: p. 97-133.

241. Peterson, R.L., 1974. A review of the general life history of the moose. *Naturaliste Canadien* 101: 9-21.

242. Poelker, R.J. and H.D. Hartwell, 1973. Black Bear of Washington: Washington State Game Department Biology.

243. Popanoe, H., 1981. *The Water Buffalo: New Prospects for an Underutilized Animal*. Washington, D. C.: National Academy Press.

244. Powell, R.A., 1981. *Martes pennanti*. *Mammalian Species* 156: 6.

245. Powell, R.A., 1982. *The Fisher: Life History, Ecology and Behavior*. Minneapolis: University of Minnesota Press.

246. Prescott, J., 1987. The status of the Thailand brow-antlered deer (*Cervus eldi siamensis*) in captivity. *Mammalia* 51: 571-577.

247. Purves, P.E. and G. Pilleri, 1978. The functional anatomy and general biology of *Pseudorca crassidens* (Owen) with a review of the hydrodynamics and acoustics in cetacea. *Investigative Cetacea* 8: 68-227.

248. Rageot, R., 1978. *Observaciones Sobre el Monito del Monte*. Silvestre: Chile Ministry Agricultura Corporacion Nacional For. Departamento Tecnical IX--Reg. Interp. --V.

249. Rahm, U., 1969. Zur Fortpflanzungsbiologie von *Tachyoryctes ruande* (Rodentia, Rhizomyidae). *Revue Suise Zoologie*. 76: 695-702.

250. Rajagopalan, P.K., 1968. Notes on the Malabar spiny dormouse, *Platacanthomys lasiurus* Blyth, 1859, with new distribution record. *Journal of Bombay Natural History Society* 65: 214-215.

251. Rajasingh, G.J., 1984. A note on the longevity and fertility of the blackbuck, *Antilope cervicapra* (Linnaeus). *Journal of Bombay Natural History Society* 80: 632-634.

252. Ralls, K., 1978. *Tragelaphus eurycerus*. *Mammalian Species* 111: 4.

253. Ramsay, M.A. and I. Stirling, 1988. Reproductive Biology and Ecology of Female Polar bears (*Ursus maritimus*). *Journal of Zoology* 214: 601-634.

254. Ray, G.C., 1981. Ross seal *Ommatophoca rossi*, in *Handbook of Marine Animals*, S.H. Ridgway and R. Harrison, Editors. London: Academic Press: p. 237-260.
255. Reeves, R.R. and S. Tracey, 1980. *Monodon monoceros. Mammalian Species* 127: 7.
256. Rice, D.W., 1978. Sperm whales, in *Marine Mammals of the Eastern North Pacific and Arctic Waters*, D. Haley, Editor. Seattle, WA: Pacific Search Press: p. 82-87.
257. Rich, M.S., 1983. The Longevity Record for *Lynx canadensis* Kerr, 1972. *Zoologische Garten N. F. Jena* 53: 365.
258. Richard, P.B., 1976. Determination de l'age et de la longevite chez le desman des Pyrenees (*Galemys pyrenaicus*). *Terre Vie* 30: 181-192.
259. Rieger, I., 1979. A review of the biology of striped hyaenas, *Hyaena hyaena* (Linne, 1758). *Saugetierk. Mitt.* 27: 81-95.
260. Rieger, I., 1981. *Hyaena hyaena. Mammalian Species* 150: 5.
261. Rigby, R.G., 1972. A study of the behaviour of caged *Antechinus stuartii. Zeitschrift Tierpsychologia* 31: 15-25.
262. Roberts, M., E. Maliniak, and M. Deal, 1984. The reproductive biology of the rock cavy, *Kerodon rupestris*, in captivity: A study of reproductive adaptation in a trophic specialist. *Mammalia* 48: 253-266.
263. Roberts, T.J., 1977. *The Mammals of Pakistan*. London: Ernest Benn.
264. Ronald, K. and P.J. Healey, 1981. Harp seal--*Phoca groenlandica*, in *Handbook of Marine Animals*, S.H. Ridgway and R. Harrison, Editors. London: Academic Press: p. 55-87.
265. Rood, J.P., 1972. Ecological and behavioural comparisons of three genera of Argentine cavies. *Animal Behavior Monograph* 5: 1-83.
266. Rosevear, D.R., 1965. *The Bats of West Africa*. London: Eyre and Spottiswoode Ltd.
267. Rosevear, D.R., 1969. *The Rodents of West Africa*. London: British Mus. (Nature Hist.).
268. Rosevear, D.R., 1974. *The Carnivores of West Africa*. London: British Mus. (Nature Hist.).
269. Ross, C., 1991. Life history pattern of New World monkeys. *International Journal of Primatology* 12: 481-502.
270. Ross, G.J.B., 1977. The taxonomy of bottlenosed dolphins *Tursiops* species in South African waters, with notes on their biology. *Annals Cape Providence Museum (Natural History)* 11: 135-194.
271. Rotterman, L.M. and T. Simon-Jackson, 1988. *Sea Otter, Enhydra lutris*. Washington, D. C.: Marine Mammal Commission.
272. Rusch, D.A. and W.G. Reeder, 1978. Population ecology of Alberta red squirrels. *Ecology* 59: 400-420.
273. Sadleir, R.M.F.S., 1987. Reproduction of female cervids, in *Biology and Management of the Cervidae*, M.W. C, Editor. Washington, D.C.: Smithsonian Institute Press: p. 123-144.
274. Savic, I., 1973. Ecology of the species *Spalax leucodon* Nordm. in Yugoslavia. *Zbornik za Prir. Nauke. Matice Srpske* 44: 66-70.
275. Schaller, G.B., 1972. *The Serengeti Lion: A Study of Predator-Prey Relations*. Chicago: University of Chicago Press.
276. Schauenberg, P., 1978. Note sur le rat de Cuming *Phloemys cumingi* Waterhouse 1839 (Rodentia, Phloeomyidae). *Revue Suisse Zoologie* 85: 341-347.

277. Schmidt, C.R., 1975. Captive breeding of the vicuna, in *Breeding Endangered Species in Captivity*, R.D. Martin, Editor. London: Academic Press: p. 271-283.

278. Schowalter, D.B., L.D. Harder, and B.H. Treichel, 1978. Age composition of some Vespertilionid bats as determined by dental anuli. *Canadian Journal of Zoology* 56: 355-358.

279. Schusterman, R.J., 1981. Steller sea lion--*Eumetopias jubatus:* Vol. 1. The walrus sea lions, fur seals, and sea otter, in *Handbook of Marine Mammals*, S.H. Ridgway and R. Harrison, Editors. London: Academic Press: p. 119-141.

280. Schwartz, C.W. and E.R. Schwartz, 1959. *The Wild Animals of Missouri*: University Missouri Press.

281. Seargent, D.E., D.J.S. Aubin, and J.R. Geraci, 1980. Life history and northwest Atlantic status of the Atlantic white-sided dolphin, *Lagenorhynchus acutus. Cetology* 37: 12.

282. Sergeant, D.E., 1973. Biology of white whales (*Delphinapterus leucas*) in western Hudson Bay. *Journal of Fisheries Research Board Canada* 30: 1065-1090.

283. Sergeant, D.E., D.K. Caldwell, and M.C. Caldwell, 1973. Age, growth, and maturity of bottlenosed dolphin (*Tursiops truncatus*) from northeast Florida. *Journal of Fisheries Research Board of Canada* 30: 1009-1011.

284. Shane, S., R.S. Wells, and B. Wursig, 1986. Ecology, Behavior and Social Organization of the Bottlenose Dolphin: A Review. *Society for Marine Mammalogy* 2: 34-63.

285. Sherman, P.W. and M.L. Morton, 1984. Demography of Belding's ground squirrels. *Ecology* 65: 1617-1628.

286. Shield, J., 1968. Reproduction of the Quokka *Setonix brachyurus*, in captivity. *Journal of Zoology* 155: 427-444.

287. Shoshani, J. and J.F. Eisenberg, 1982. *Elephas maximus. Mammalian Species* 182: 8.

288. Shoshani, J., C.A. Goldman, and J.G.M. Thewissen, 1988. *Orycteropus afer. Mammalian Species* 300: 8.

289. Simons, L.S., 1984. Seasonality of reproduction and dentinal structures in the harbor porpoise (*Phocena phocena*) of the North Pacific. *Journal of Mammalogy* 65: 491-495.

290. Siniff, D.B. and S. Stone, 1985. *The Role of the Leopard Seal in the Tropho-Dynamics of the Antarctic Marine Ecosystem.* Berlin: Springer-Verlag.

291. Skinner, J.D., J.H.M. VanZyl, and J.A.H. VanHeerden, 1973. The effect of season on reproduction in the black wildebeest and red hartebeest in South Africa. *Journal of Reproduction and Fertility Supplement* 19: 101-110.

292. Slobodyan, A.A., 1976. The European brown bear in the Carpathians, in *Bears--Their Biology and Management*, M.R. Pelton, J.W. Lentfer, and G.E. Folk, Editors. International Union Conservation Nature Publ.: p. 313-319.

293. Smielowski, J., 1985. Longevity of carnivores from South America. *Zool. Garten N. F. Jena* 55: 177.

294. Smith, A., 1984. Demographic consequences of reproduction, dispersal and social interaction in a population of Leadbeater's possum, in *Possums and Gliders*, A. Smith and I. Hume, Editors. Norton, New South Whales: Surrey Beatty & Sons: p. 359-373.

295. Smithers, R.H.N., 1971. *The Mammals of Botswana*: Trustees National Museums and Monuments Rhodesia Museum Mem.

296. Smithers, R.H.N., 1983. *The Mammals of the Southern African Subregion*: University Pretoria.

297. Smolen, M.J., H.H. Genoways, and R.J. Baker, 1980. Demographic and reproductive parameters of the yellow-cheeked pocket gopher (*Pappogeomys castanops*). *Journal of Mammalogy* 61: 224-236.

298. Snyder, R.L. and S.C. Moore, 1968. Longevity of captured mammals in Philadelphia Zoo. *International Zoo Yearbook* 8: 175-183.

299. Spinage, C.A., 1982. *A territorial antelope: The Uganda waterbuck*. London: Academic Press: p. 334.

300. Stirling, I., W. Calvert, and D. Andriashek, 1980. Population ecology studies of the polar bear in the area of southeastern Baffin Island. *Canadian Wildlife Service Occasional Papers* 44: 33.

301. Stodart, E., 1977. Breeding and behaviour of Australian bandicoots, in *The Biology of Marsupials*, B. Stonehouse and D. Gilmore, Editors. Baltimore: University Park Press: p. 179-191.

302. Strahan, R., 1975. Status and husbandry of Australian monotremes and marsupials, in *Breeding Endangered Species in Captivity*, R.D. Martin, Editor. London: Academic press: p. 171-182.

303. Strahan, R., ed. 1983. *The Australian Museum Complete Book of Australian Mammals*. ed. Series. Strahan, R. London: Angus & Robertson Publ.

304. Stroganov, S.U., 1969. *Carnivorous Mammals of Siberia*. Jerusalem: Israel Progr. Sci. Transl.

305. Strong, J.T., 1988. Status of the Narwhal, *Monodon monoceros*, in Canada. *Canadian Field Naturalist* 102: 391-398.

306. Sugimura, M., Y. Suzuki, I. Kita, Y. Ide, S. Kodera, and M. Yoshizawa, 1983. Prenatal development of Japanese serows, *Capricornis crispus*, and reproduction in females. *Journal of Mammalogy* 64: 302-304.

307. Sunquist, M.E., 1982. Movements and habitat use of a sloth bear. *Mammalia* 46: 545-547.

308. Taylor, P.J., J.U.M. Jarvis, T.M. Crowe, and K.C. Davies, 1985. Age determination in the cape mole rat *Georychus capensis*. *South African Journal of Zoology* 20: 261-267.

309. Thomas, J., V. Pastukhov, R. Elsner, and E. Petrov, 1982. *Phoca sibirica*. *Mammalian Species* : 6.

310. Tiba, T., M. Sato, T. Hirano, I. Kita, M. Sugimura, and Y. Suzuki, 1988. An annual rhythm in reproductive activities and sexual maturation in male Japanese serows (*Capricornis crispus*). *Zeitschrift fur Saugetierkunde* 53: 178-187.

311. Treus, V.D. and N.V. Lobanov, 1971. A climatisation and domestication of the eland *Taurotragus oryx* at Askanya-Nova Zoo. *International Zoo Yearbook* 11: 147-156.

312. Tullar, B.F.J., 1983. A long-lived white-tailed deer. *New York Fish and Game Journal* 30: 119.

313. Tuttle, M.D. and D. Stevenson, 1982. Growth and survival of bats, in *Ecology of Bats*, T.H. Kunz, Editor. New York: Plenum Press: p. 105-150.

314. Tyndale-Biscoe, C.H. and R.B. MacKenzie, 1976. Reproduction in *Didelphis marsupialis* and *D. albiventris* in Colombia. *Journal of Mammalogy* 57: 248-265.

315. USNMFS, 1978. The Marine Mammal Protection Act of 1972: Annual Report 1977-78: . Washington, D. C.: U. S. National Marine Fisheries Service.

316. Uspenski, S.M. and S.E. Belikov, 1976. Research on the polar bear in the USSR, in *Bears--Their Biology and Management*, M.R. Pelton, J.W. Lentfer, and G.E. Folk, Editors. International Union Conservation Nature Publ.: p. 321-323.

317. VanBallenberghe, V., A.W. Erickson, and D. Byman, 1975. Ecology of the Timber Wolf in Northeastern Minnesota. *Wildlife Monographs* 43: 43.

318. VanBallenberghe, V. and L.D. Mech, 1975. Weights, growth and survival of timber wolf pups in Minnesota. *Journal of Mammalogy* 56.

319. VanHeerden, J. and F. Kuhn, 1985. Reproduction in captive hunting dogs *Lycaon pictus. South African Journal of Wildlife Research* 15: 80-84.

320. VanHorn, R.N. and G.G. Eaton, 1979. Reproductive physiology and behavior in prosimians, in *The Study of Prosimian Behavior*, G.A. Doyle and R.D. Martin, Editors. New York: Academic Press: p. 79-122.

321. VanLawick, H. and J.V. Lawick-Goodall, 1971. *Innocent Killers*. Boston: Houghton Mifflen.

322. VanRompaey, H., 1978. Longevity of a banded mongoose (*Mungos mungo gmelin*) in captivity. *International Zoo News* 25: 32-33.

323. Wackernagel, H., 1968. A note on breeding the serval cat *Felis serval* at Basel Zoo. *International Zoo Yearbook* 8: 46-47.

324. Walley, H.D. and W.L. Jarvis, 1971. Longevity record for *Pipistrellus subflavus. Transactions Illinois Academy Science* 64: 305.

325. Wang, F., 1985. Preliminary study of the complex-toothed flying squirrel, in *Contemporary Mammalogy in China and Japan*, T. Kawamichi, Editor. Osaka: Mamm. Soc. Japan: p. 67-69.

326. Waring, G.H., 1983. *Horse Behavior*. Park Ridge, NJ: Noyes Publ.

327. Warneke, R.M. and P.D. Shaughnessy, 1985. *Arctocephalus pusillus*, the South African and Australian fur seal: taxonomy, evolution, biogeography, and life history, in *Study of Sea Mammals in South Lattitudes*, J.K. Ling and M.M. Bryden, Editors. Sydney: Southern Australia Museum: p. 53-77.

328. Watkins, L.C., 1972. *Nycticeius humeralis. Mammalian Species* : 4.

329. Weir, B.J., 1974. Reproductive characteristics of Hystricomorph rodents. *Symposium Zoological Society London* 34:265-301.

330. Wemmer, C.M., 1977. Comparative ethology of the large spotted genet *Genetta tigrina* and some related viverrids. *Contributions to Zoology* : 93.

331. Wiloughby, D.P., 1974. *The Empire of Equus*. London: Thomas Yoseloff.

332. Woolley, P., 1982. Observations on the feeding and reproductive status of captive feather-tailed possums, *Distoechurus pennatus* (Marsupialia: Burramyidae). *Australian Mammalogy* 5: 285-287.

333. Wursig, B. and M. Wursig, 1980. Behavior and ecology of the dusky dolphin, *Lagenorhynchus obscurus*, in the South Atlantic. *Fishery Bulletin* 77: 871-890.

334. Yager, R.H. and C.B. Frank, 1972. The nine-banded armadillo for medical research. *Institute Laboratory Animal Research News* 15: 4-5.

335. Yamamoto, S., 1967. Breeding Japanese serows *Capricornis crispus* in captivity. *International Zoo Yearbook* 7: 174-174.

336. Ziegler, A.C., 1977. Evolution of New Guinea's marsupial fauna in response to a forested environment, in *The Biology of Marsupials*, B. Stonehouse and D. Gilmore, Editors. Baltimore: University Park Press: p. 117-138.

Birds

Flight, the major avian adaptation, is thought to have contributed to birds' rapid speciation and successful adaptations to habitats world wide. Class Aves includes more than 8500 species of birds and is divided into two superorders, the Palaeognathae (e.g., ratites and tinamous) and the Neognathae (e.g., all other modern birds). Classification of birds is fluid. Bird life spans vary from the maximal life span of 1.5 years in the Greater Prairie Chicken (ref. Tymphanuchus cupido) to the 58 year life span of the RoyalAlbatross (Diomedea chrysostoma. The vast majority of birds are socially monogamous and a sizable subset maintain pair bonds over many annual reproductive attempts.

Paleognathae

STRUTHIONIFORMES (ostriches, emus, cassowaries, kiwis)
Struthioniformes include the ostrich of Africa as well as emus, cassowaries and kiwis. Struthioniformes arose in the Miocene, late in the evolution of most modern birds (Proctor and Lynch 1993). With some weighing in excess of 150 kilograms and standing over two meters, Struthioniformes are the largest living birds (Pettingill 1985, Welty 1982). Unique features include a flat sternum, heavily-muscled legs, and two to four short toes. Birds of this order have sparse feathering on the head, neck, or legs and relatively small brains. They hatch precocial young from eggs which weigh roughly three pounds, measure 7 by 5.5 inches, and comprise 1.7 percent of the adult's body weight (Pettingill 1985: 295). Ostriches are polygynous (e.g., males mate with two or more females), a characteristic which has been positively associated with birds living in open habitats (Verner 1964).

Kiwis are flightless, terrestrial birds with four toes and hair-like feathers. Kiwis (as well as some sea birds and nocturnal birds of prey) have a higher body temperature at night than that during the day (Gill 1995) and experience greater body temperature fluctuations and lower resting temperatures than other birds. Kiwis have a superior sense of smell which may be attributable to their elongated bill with nostrils at the tip, large nasal cavities, and mammal-like olfactory system (Pettingill 1985:104). Kiwi eggs are large relative to total body mass; egg mass is approximately a quarter of the bird's total body weight (Pettingill 1985, 295). The

Brown Kiwi's 75-80 day incubation is one of the longest incubation periods among birds (Frith 1962).

RHEIFORMES (rheas)

The two species of rheas are flightless and distributed through the neotropics. The earliest fossil record for Rheiformes appeared in the Eocene epoch (37 to 53 million years ago, Proctor and Lynch 1993). Like the Struthioniformes, rheas have a flat sternum, heavy legs, and precocial young. However, they are smaller (1.5 meters) and originated in South America.

TINAMIFORMES (tinamous)

Tinamous are distributed throughout the Neotropics, and include 47 species in 9 genera. The earliest tinamous appeared in the Pliocene epoch between two and several million years ago (Proctor and Lynch 1993). Tinamiformes are poor fliers though they have functional wings. The young are precocial (Gill 1995:619). Unique to Tinamiformes and ratites (Rheiformes, Struthioniformes, Casuariiformes, and Dinornithiformes) is a configuration of nasal bones known as the paleognathous palate. The Solitary Tinamou has lived for as long as 15 years in captivity.

Neognathae

SPHENISCIFORMES (penguins)

There are 17 species in one family of penguins found between the Southern Hemisphere and the Galapagos Islands. They are piscivorous marine divers with webbed feet and modified wings for swimming (Welty 1985). Sphenisciformes are characterized by a large keel and small, dense, waterproof feathers. They lack brood patches, and instead monogamous pairs incubate 1 to 2 eggs (and subsequently brood young) on top of their feet (Gill 1995:392, 625). Penguins are thought to have evolved from Procellariform-like birds with whom they share tubular nostril construction (Simpson 1946, Sibley and Ahlquist 1990, Ho et al. 1976), and similar courting and feeding behaviors (Gill 1995:625).

GAVIIFORMES (loons)

The four species of loons can be found in North America and Eurasia. They breed along lakes and rivers and winter along the coast. Loons are webbed-toed divers with short legs well back near the tail. Loons dive to depths of 75 meters for periods up to 8 minutes at a time to forage for fish, frogs, and aquatic invertebrates (Gill

1995:649). Gaviiformes are perenially monogamous and produce small broods of precocial young after 3 to 4 weeks of incubation (Gill 1995:649).

PODICIPEDIFORMES (grebes)

Twenty-one extant species of grebes inhabit lakes, ponds, and marshes (Udvardy and Farrand 1994). They are an old order (approximately 70 million years). They are pandemic, medium-sized diving birds with short, posterior-positioned legs and lobed toes adapted for swimming but resulting in limited land movement. Feeding on fish, insects and other aquatic animals, grebes digest their own feathers to trap fish bones for later regurgitation or delayed digestion (Gill 1995:623). Grebes build platform nests over water using aquatic weeds. They winter along coastlines. Most grebes are black and white with little sexual dimorphism.

PROCELLARIIFORMES (tube-nosed sea birds)

Procellariiformes (albatrosses, shearwaters and petrels) are distributed worldwide. Some of the oldest fossils date back more than 65 million years (Proctor and Lynch 1993). These birds have tubular nostrils and a keen sense of smell, a hooked beak, oily feathers, long narrow wings, and webbed feet (Gill 1995:627; Udvardy and Farrand 1994). They vary in size from the six inch Least Storm Petrel to the Wandering Albatross with an almost twelve foot wingspan (Gill 1995:627). The typically grey, black, brown, and white-colored Procellariiformes are particularly noteworthy for their near-effortless ability to soar for long periods of time.

Residing in the oceans for most of the non-breeding season, most Procellariiformes are perennially monogamous. Procellariiformes feed on marine animals (e.g., fish, squid, crustaceans) or plankton on the surface of the water. Shearwaters and Petrels live on the oceans during the non-breeding season, breed on islands and coastlines, and often winter in North America (Udvardy and Farrand 1994). Some are sexually dimorphic and breed in burrows along the coast and on islands (Gill 1995:627; Udvardy and Farrand 1994). Most Petrel pair bondings last for one season (65%), with 58% of bonds attributable to loss of a partner (Warham 1990, 204). Divorces appear to be rare among petrels.

Most albatrosses breed on pelagic island beaches or along remote coastal regions. They are perennially monogamous and lay one egg per clutch. Incubation and fledging take 190 to 365 days; species with the longest nestling periods reproduce only every other year or less. Age-dependent reproductive rate of sea birds increases with age (Wooller et al. 1992), while death rates may not rise geometrically over time.

PELECANIFORMES (pelicans, gannets, and cormorants)

Pelecaniformes include pelicans (8 species), tropicbirds (3 species), boobies and gannets (9 species), cormorants (38 species), frigatebirds (5 species), and snakebirds (4 species). Fossil records indicate that pelicans first appeared 26 to 37 million years ago (Proctor and Lynch 1993). The order is primarily distributed through the Tropics and Subtropics world-wide. Most Pelecaniformes are medium to large bodied, fish and squid-eating birds. They have webbed feet with four toes, a long bill, poorly or undeveloped nostrils, and a gular sac (throat pouch). Cormorants, boobies, and gannets lack external nostril openings and breathe through the mouth (Gill 1995:629). Feeding techniques vary from diving and piercing or pursuing fish in the water, to air-borne diving and scooping activities. With the exception of tropicbirds, Pelecaniformes are born without down. Monogamous pairs nest in colonies on the ground, on cliffs, or in trees, laying one to three eggs per clutch.

CICONIIFORMES (wading birds, herons and storks)

Ciconiiformes includes 6 families with 119 species of herons, egrets, bitterns, storks, ibises, spoonbills, and flamingos. Storks first appeared 37 to 53 million years ago, while ibises and herons first 53 to 65 million years ago (Proctor and Lynch 1993). Stork habitats include tropical and temperate marshes of Eurasia (Udvardy and Farrand 1994). Sixty-five species of medium-sized bitterns and herons are found worldwide. Unique to Herons and Shoebills are powderdowns, disintegrating feathers which condition the remainder of their plumage, and unique to Ibises and Shoebills are grooves on the bill which run from the nostril to the bill tip (Gill 1995:635).

Ranging in size from the foot-tall Least Bitterns to the five foot Goliath Herons, Ciconiiformes are waders with long necks, long legs, and (except for Flamingos) un-webbed toes. They feed predominantly on fish, frogs and aquatic or terrestrial invertebrates. Nesting singly or in colonies, Ciconiiformes build stick nests in marshes or along shores (Welty 1984); however, storks, herons, and ibises build their nests in trees and bushes (Udvardy and Farrand 1994) and sometimes rooftops.

ANSERIFORMES (ducks, geese and swans)

Anseriformes includes over 160 species in four families, the Anatidae (swans, geese and typical ducks), Dendrocygnidae (Whistling ducks), Anseranatidae (Magpie goose), and Anhimidae (screamers). Anseriformes are relatively modern, having appeared 26 to 37 million years ago (Gill 1995:Proctor and Lynch 1993).

Characteristics common to this order include depressed bills with filtering "teeth" for surface feeders and stronger, broader bills for deeper divers. All have short legs, webbed feet, oily feathers, unspotted eggs and hatch downy, precocial chicks. While most waterfowl are monogamous, male Anseriformes do not assist in chick-rearing (Gill 1995:631). Geese and swans are sexually monomorphic and perennially monogamous while ducks are sexually dichromatic and seasonally monogamous. The young of some geese remain with the parents for two years resulting in overlapping broods.

Although most Anseriformes are aquatic, geese forage primarily in terrestrial habitat (Evardy, 1994) while ducks forage primarily in water. Across the species, diets include plant parts as well as insects, fish, and snails and other aquatic invertebrates. Body lengths range from 12 to 72 inches (Pettingill 1985). Anseriformes may be related to the early sea birds or to Galliformes (Johnsgard 1968, Prager and Wilson 1976).

FALCONIFORMES (birds of prey)

The predatory Falconiformes include the diurnal vultures, eagles, hawks, kites, harriers, osprey, caracaras, and falcons. Wingspans range from 13 centimeters in the Philippine Falconet to 3 meters in the Andean Condor (Gill 1995:637). Eagles and hawks appear to have evolved during the Eocene, while falcons appeared in fossil record more recently during the Miocene (Proctor and Lynch 1993). This order is characterized by strong, sharp, hooked beaks with a fleshy cere at the bill base and sharply-pointed talons. Reversed sexual dimorphism (females larger than males) is common. Most Falconiformes breed in trees; they lay three to five eggs per clutch (fewer in larger species) and hatch downy young whose eyes are already open. After leaving the nest, young continue to receive food from parents for some months (Gill 1995:637).

Accipitridae includes 240 species of hawks and eagles distributed worldwide. These raptors generally take ground or aquatic prey and build nests in trees or on pinnacles. The Falconidae include 63 species of falcons and caracaras that inhabit open country and take prey on the wing. Falcons nest in cliff crevices, hollow trees, and on the ground. Recent analysis confirms the monophyly of Falconiformes (Griffiths 1994) although some researchers speculate that New World vultures are more closely related to the Ciconiiformes (Ligon 1967, Sibley and Ahlquist 1990).

GALLIFORMES (fowl-like birds, gallinaceous birds)

There are 263 species of Galliformes in six families. They are found world-wide (except Antarctica and southern South America). Galliformes include curassows, grouse, ptarmigan, quails, partridge, pheasants, peafowl, and turkeys. Galliformes are strong runners with heavy feet and strong breast muscles. They nest on the ground and are subject to high rates of predation They are herbivorous and insectivorous, often gregarious, and nest on the ground. Their precocial young are commonly hatched in large clutches that band into larger groups. Most species are polygynous (some form mating leks) and exhibit both body size and chromatic sexual dimorphism.

GRUIFORMES (cranes, rails, and coots)

The order Gruiformes includes 14 families with 170 species of prairie and marsh-dwelling birds, of which 80% are rails and coots. They date back 37 to 53 million years (Proctor and Lynch 1993). Gruiformes lack a crop, and possess short, rounded wings. Birds of this order are terrestrial, aquatic, or marsh-dwelling, and most have slightly-webbed or unwebbed toes. With the exception of the cranes and bustards, the young tend to be precocial. Some members of this order are polyandrous (females compete for and mate with many males).

Rails have laterally compressed bodies adept for movement among reeds and cattails, and large feet with long toes. Coots are adept open-water swimmers with lobed feet. Gallinules are more brightly colored than the more cryptic rails and coots (Udvardy and Farrand 1994).

CHARADRIIFORMES (shorebirds and gulls)

Order Charadriiformes is comprised of 18 families (more than 300 species) of shore birds, such as oystercatchers, plovers, snipes, sandpipers, avocets, gulls, terns, and auks. This group is sometimes included as a suborder of the Ciconiiformes (Sibley and Ahlquist 1990) but we consider them separately. Shorebirds are an old group, dating back more than 65 million years (Proctor and Lynch 1993). The toes of these birds are often webbed, their plumage compact, and they tend to be strong fliers. Dietary habits of Charadriiformes range from probing and plucking insects from mud, scavenging on shorelines, skimming the oceans, and stealing food from other birds, to deeper diving activities in search of marine fish. Many species are gregarious, however nesting is established singly and on the ground. Young are precocial.

More gulls than terns may be found in cooler climates (Udvardy and Farrand 1994). Terns which primarily feed on fish are often characterized with black caps and straight bills while gulls may be characterized with slightly hooked bills, semi-precocial young, and white, black, and grey markings.

COLUMBIFORMES (pigeons and doves)

Worldwide, Columbiformes include more than 300 small, medium, and large size pigeon and dove species as well as the extinct dodo. Columbiformes have narrow bills, cere at the base of the bill, short legs with reticular scales, small neck, and a small head. Adults lay clutches of two eggs from which altricial young are hatched. Columbiformes young are fed a rich substance produced from the lining of the crop known as "pigeon milk" (Welty 1985) by both males and females. Pigeons are unique in their ability to drink water by sucking rather than tilting and tossing water back with their heads (Gill 1995:651).

PSITTACIFORMES (parrots)

Psittaciformes includes approximately 350 species of tropical parrots, macaws, and lories and comprises the most endangered group with high rates of threatened species (Gill 1995:653). Psittaciformes are characterized by a strong, narrow, hooked beak, an upper mandible which is hinged movably to the skill, and green, colorful, or black plumage. Size varies from 7 to 90 centimeters, their tongues are fleshy, and their toes are positioned two in the front and two in the back. Parrots and macaws are predisposed to nesting in holes, and are gregarious.

CUCULIFORMES (cuckoos)

Cuculiformes encompasses two families -- Musophagidae (Turacos) and Cuculidae (cuckoos and roadrunners) that are mostly native to the tropics. Members of Family Cuculidae include cuckoos, coucals, roadrunners, and anises. Old World cuckoos are brood parasites – laying their eggs in the nests of other birds. American cuckoos lay eggs in their own nests (Udvardy and Farrand 1994). Cuculiformes date from 26 to 37 million years ago (Proctor and Lynch 1993). This order is not closely related to any other extant order.

STRIGIFORMES (owls)

Strigiformes include over 140 species of effective, silent, nocturnal predators. They have large more forward facing eyes and bony plates of cylindrical construction useful for telescopic vision (Gill 1995:661). In addition to keen hearing, owls have facial feather plates that amplify sound. Soft, dense plumage allows for silent flight. With strong hooked beaks, strong feet, sharp talons, and feathered legs, owls feed mainly on small rodents and shrews as well as on insects, frogs, fish, reptiles, and other birds (Gill 1995:661). Generally, owls use abandoned nests of other birds, burrows, or natural cavities, laying clutch sizes according to resource availability. Strigiformes is one of the older orders with origins between 53 and 65 million years ago (Proctor and Lynch 1993).

CAPRIMULGIFORMES (nightjars and goatsuckers)

There are 98 species of Caprimulgiformes in five families including Nightjars, Oilbirds, and Goatsuckers. They are small to medium sized insectivores. The wide mouths of Caprimulgiformes have insect-netting bristles which improve their ability to successfully hunt for insects at twilight and which protect their eyes from damage related to in-flight insect collisions. They lay one or two eggs directly onto the ground, into tree crevices, or on cave ledges (Gill 1995:663).

APODIFORMES (swifts and hummingbirds)

There are two families of Apodiformes, the Hemiprocnidae (swifts) and the Trochilidae (hummingbirds). Apodiformes have pointed wings, short thick humeri, and unfeathered oilgland, and lack of a crop in adults. The Trochilidae include more than 100 species of hummingbirds that range from 2 to 20 grams and are sexually dichromatic with brilliant, often iridescent, plumage. They are primarily nectivores with some insectivory and are able to torpor. Hummingbirds are polygynous New World birds and females raise their young without paternal assistance. The 74 species of swifts range from 9 to 150 grams and are not sexually dimorphic. They catch insects on the wing and some may echolocate. This order may be paraphyletic (Sibley and Ahlquist 1990).

COLIIFORMES (mousebirds)

Coliidae, the one family of Coliiformes includes six species of mousebirds and colies all native to sub-Saharan Africa. Coliiformes originated 7 to 26 million years ago (Proctor and Lynch 1993).

The characteristically gregarious Coliiformes are small birds with crests, long tails, and opposable first and fourth toes which are used for climbing through brush. They build platform-like nests in their savannah, woodland, and brushy habitats, and feed on fruits, berries, buds and flowers (Gill 1995:655). Clutches of 2 to 4 eggs are incubated, at least in some species, by both sexes. Young hatch naked and altricial.

TROGONIFORMES (trogons)

There are 34 species in the two families of Trogoniformes. All trogons are tropical and subtropical. They have short broad bills and wings, weak feet with the first two toes pointed backward and the third and fourth toes pointed forward, long tails, and iridescent, dense and typically green plumage. Trogoniformes feed on insects and fruit. They nest in cavities where they hatch blind and naked chicks (Gill 1995:667). They may be distantly related to Coraciiformes (Maurer and Raikow 1981, Sibley and Ahlquist 1990).

CORACIIFORMES (kingfishers, rollers, bee-eaters, hoopoes)

Nine families of small to medium sized birds comprise the Coraciiformes. There are 94 species of kingfishers (Alcedinidae), twelve rollers (Coraciidae), 26 bee-eaters (Meropidae), and 56 species of hornbills (Bucerotidae) in this order that originated 53 to 65 million years ago (Proctor and Lynch 1993). Coraciiformes have a strong bill, toes with the third and fourth toes joined at the base, and brightly colored plumage. Their omnivorous diet includes fruits, arthropods, fish and invertebrate aquatic life, and worms. They hunt using several strategies (e.g., sit and wait, probing into crevices , and feeding in flight). These birds nest in cavities, raising altricial young. Social organization varies from perenially monogamous (hornbills) to seasonal monogamy (kingfishers) to cooperative-breeding (kookaburras, bee-eaters, southern hornbills). Coraciiformes may not be monophyletic (Cracraft 1981; Maurer and Raikow 1981).

PICIFORMES (woodpeckers, toucans, jacamars and puffbirds)

Piciformes are comprised of the tropical and subtropical toucans, honeguides, and barbets (Ramphastidae, Indicatoridae, and Capitonidae), the tropical, subtropical, and temperate woodpeckers (Picidae), and neotropical, jacamars and puffbirds (Galbulidae, Bucconidae). Piciformes have highly specialized bills for feeding on foods such as worms, caterpillars, and insects found beneath the bark or in seeds of trees; however, Toucans also eat the eggs and young of other birds. Stiffened tail feathers facilitate upright perching on tree trunks and branches (Gill 1995:671). Little sexual dimorphism is apparent in the Piciformes; they nest in cavities or holes and bear altricial young. Monogamy and biparental care is common, however, some species are cooperative breeders (aracaris) and some are nest parasites (honeyguides).

PASSERIFORMES (perching or song birds)

Fifty percent of 9672 known bird species are Passeriformes. Due to convergent evolution, subdivision of the order into smaller groups is difficult. Using DNA comparisons and the same criteria as used with non-passerines, Sibley and Ahlquist (1990) divided Passeriformes into 45 families. Passeriform feet are adapted to perching. These birds have relatively large brains and exhibit high capacity for learning vocalizations (Gill 1995:675). With a few exceptions, Passeriformes feed on seeds, fruit or nectar, or insects. All forms of mating system and social organization are found; however, in all, young are altricial. Crows and jays are evidenced in fossil records as of the Miocene Period while sparrows, swallows, and nuthatches appeared more recently (Proctor and Lynch 1993).

References Cited

Cracraft, J., 1981. "Toward a phylogenetic classification of the recent birds of the world." *Auk* 98: 681-714.

Frith, H. J., 1962. *The Mallee-fowl*. Sydney, Angus and Robertson.

Gill, F. B., 1995. *Ornithology*. New York, W. H. Freeman and Company.

Griffiths, C. S., 1994. "Monophyly of the Falconiformes based on syringeal morphology." *Auk* 111(787-805).

Johnsgard, P. A., 1968. *Waterfowl: Their Biology and Natural History*. Lincoln, University of Nebraska Press.

Ligon, J. D., 1967. *Relationship of the cathartid vultures*. Occasional Papers, Ann Arbor, University of Michigan Museum of Zoology.

Mauere, D. R. and R. J. Raikow, 1981. "Appendicular myology, phylogeny, and classification of the avian order Coraciiformes (including Trogoniformes)." *Annals of the Carnegie Museum* 50: 417-434.

Pettingill, O. S. J., 1985. *Ornithology in Laboratory and Field*. Minneapolis, Burgess Publishing Company.

Prager, E. M. and A. C. Wilson, 1976. "Congruency of phylogenies derived from different proteins." *Journal of Molecular Evolution* 9: 45-57.

Proctor, N. S. and P. J. Lynch, 1993. *Manual of Ornithology: Avian Structure and Function*. New Haven, Yale University Press.

Sibley, C. G. and J. E. Ahlquist, 1990. *Phylogeny and Classification of Birds: A Study in Molecular Evolution*. New Haven, Yale University Press.

Simpson, G. G., 1946. "Fossil penguins." *Bulletin of the American Museum of Natural History* 87: 1-100.

Udvardy, M. and J. Farrand, 1994. *National Audubon Society Field Guide to North American Birds*. New York, Alfred Knopf.

Verner, J., 1964. "Evolution of polygamy in the long-billed marsh wren." *Evolution* 18: 252-261.

Warham, J., 1990. *The Petrels: Their Ecology and Breeding Systems*. San Diego, Academic Press.

Welty, J., 1982. *The Life of Birds*. Philadelphia, W. B. Saunders.

Wooller, R. D., J. S. Bradley, et al., 1992. "Long-term population studies of seabirds." *Trends in Ecology and Evolution* 7: 111- 114.

Table 2. Record Life Spans (years) of Birds

((+) indicates that individual was still alive at the time that life span was recorded)

Order/Family	Genus/Species	Common Name	Wild.	Capt.	M/F	Reference
Anseriformes						
Anatidae	Aix sponsa	Wood duck	17.6		m	[29]
	Aix sponsa	Wood duck	22.5		m	[78]
	Alopochen aegyptiacus	Egyptian goose		25.5	x	[50]
	Anas acuta	Common pintail	20.5		m	[29]
	Anas acuta	Northern pintail	21.3		m	[78]
	Anas americana	American wigeon	21.3		m	[29]
	Anas americana	American wigeon	9		x	[37]
	Anas clypeata	Northern shoveler	16.6		m	[29]
	Anas crecca	Green-winged teal	20.5		x	[29]
	Anas crecca	Green-winged teal	20.3		f	[29]
	Anas cyanoptera	Cinnamon teal	12.9		m	[29]
	Anas diazi	Mexican duck	5.5		m	[29]
	Anas discors	Blue-winged teal	17.4		m	[29]
	Anas discors	Blue-winged teal	22.3		f	[78]
	Anas fulvigula	Mottled duck	13.4		m	[29]
	Anas fulvigula	Mottled duck (+)	13	20	x	[132]
	Anas laysanensis	Laysan teal	11.8		f	[29]
	Anas penelope	Eurasian wigeon	8.6		m	[29]
	Anas penelope	Eurasian wigeon		6.4	x	[50]
	Anas platyrhynchos	Mallard	23.4		m	[29]
	Anas platyrhynchos	Mallard	29		x	[37]
	Anas platyrhynchos	Mallard	20.5		x	[90]
	Anas rubripes	American black duck	26.4		m	[29]
	Anas strepera	Gadwall	17.7		m	[29]
	Anas strepera	Gadwall		5.4	x	[50]
	Anas strepera	Gadwall	19.5		m	[78]
	Anas wyvilliana	Hawaiian duck	3.7		f	[78]
	Anser albifrons	Greater white-fronted goose	20.3		f	[29]
	Anser albifrons	White-fronted goose		27.8	x	[50]
	Anser albifrons	Greater white-fronted goose	22.3		m	[78]
	Anser albifrons	Greater white-fronted goose	22.3	47	f	[116]
	Anser brachyrhynchus	Pink-footed goose	22		x	[37]
	Anser caerulescens	Snow goose	26.6		x	[29]
	Anser caerulescens	Blue goose	19.5		x	[29]
	Anser caerulescens	Snow (blue) goose	24.5		x	[78]
	Anser canagica	Emperor goose	9		x	[17]
	Anser canagica	Emperor goose	6.3		f	[29]
	Anser domesticus	Domestic goose		31	x	[50]
	Anser fabalis	Bean goose		23	x	[50]
	Anser rossii	Ross' goose	19.6		m	[29]
	Anser rossii	Ross' goose	21.5		m	[78] [121]
	Aythya affinis	Lesser scaup	18.3		m	[29]
	Aythya americana	Redhead	21.4		x	[29]

Aythya aythya	Canvasback	18.8		m	[29]
Aythya collaris	Ring-necked duck	16.2		m	[29]
Aythya collaris	Ring-necked duck	20.4		m	[78]
Aythya marila	Greater scaup	18.3		m	[29]
Aythya valisineria	Canvasback	22.4		m	[78]
Branta bernicla	Atlantic brant	16.6		x	[29]
Branta bernicla	Brent goose		21.8	x	[50]
Branta bernicla hrota	(Atlantic) brant	21.6		f	[78]
Branta bernicla nigricans	(Black) brant	25.8		m	[78]
Branta canadensis	Canada goose	23.5		f	[29]
Branta canadensis	Canada goose	24.3		m	[78]
Branta nigricans	Black brant	19.6		f	[29]
Bucephala albeola	Bufflehead	12.5		m	[29]
Bucephala albeola	Bufflehead	14.9		m	[37]
Bucephala albeola	Bufflehead	11.5		f	[37]
Bucephala clangula	Common goldeneye	14.3		x	[29]
Bucephala clangula	Common goldeneye	12		f	[40]
Bucephala clangula	Common goldeneye	11		m	[40]
Bucephala islandica	Barrow's goldeneye	15.3		m	[29]
Cereopsis novaehollandiae	Cape barren goose		28.1	x	[50]
Clangula hyemalis	Oldsquaw	15.3		m	[29]
Clangula hyemalis	Oldsquaw	15.6		f	[78]
Cygnus buccinator	Trumpeter swan	23.8		m	[29]
Cygnus buccinator	Trumpeter swan (+)		29.5		[50]
Cygnus buccinator	Trumpeter swan	24	32.5	x	[83] [76]
Cygnus c. columbianus	Tundra (whistling) swan (+)	19.7		f	[78]
Cygnus columbianus	Whistling swan	16.2		f	[29]
Cygnus columbianus	Tundra swan (+)	21		x	[88]
Cygnus olor	Mute swan	13.3		f	[29]
Cygnus olor	Mute swan (+)	21		x	[37]
Cygnus olor	Mute swan	26.8		m	[78]
Dendrocygna autumnalis	Black-bellied whistling duck	8.2		f	[29]
Dendrocygna bicolor	Fulvous whistling duck	6.5		m	[29]
Lophodytes cucullatus	Hooded merganser	11.3		f	[29]
Lophodytes cucullatus	Hooded merganser	10.3		x	[38]
Melanitta deglandi	White-winged scoter	11.7		f	[29]
Melanitta fusca	White-winged scoter	12		x	[37]
Melanitta fusca	White-winged scoter	15.6		f	[78]
Mergus merganser	Common merganser	13.4		f	[29]
Mergus serrator	Red-breasted merganser	5.4		x	[29]
Oxyura australis	Australian blue-billed duck		16	x	[37]
Oxyura jamaicensis	Ruddy duck	13.6		m	[29]
Plectropterus gambensis	Spur-winged goose		11.3	x	[50]
Polysticta stelleri	Stellar's eider (+)	14.3		f	[78]
Polysticta stelleri	Steller's eider	12.3		m	[29]
Somateria mollissima	Common eider	14.8		f	[29]
Somateria mollissima	Common eider	21.3		f	[78]
Somateria mollissima	Common eider	23		x	[37]
Somateria spectabilis	King eider	18.9		f	[29]
Tachyeres cinereus	Steamer duck		10.2	x	[50]

Apodiformes						
Apodidae	Aeronautes sexatilis	White-throated swift			x	[30]
	Apus apus	Swift		21	x	[141]
	Chaertura vaux	Vaux's swift	5.1		x	[30]
	Chaetura pelagica	Chimney swift	14		m	[30]
	Cypseloides niger	Black swift	6		x	[30]
Caprimulgiformes						
Caprimulgidae	Caprimulgus vociferus	Whippoorwill	4		x	[30]
	Chordeiles minor	Common nighthawk	9		f	[30]
	Phalaenoptilus nuttallii	Common poorwill (+)	3		x	[35]
Casuariiformes						
Casuariidae	Casuarius papuanus	Westermann's cassowary		26	x	[50]
	Casuarius unappendiculatus	Blyth's cassowary		22	x	[50]
Dromiceidae	Dromaius novaehollandiae	Emu (+)		10	x	[37]
	Dromaius n. hollandiae	Windsor emu		16.6	x	[50]
Charadriiformes						
Alcidae	Aethia cristatella	Crested auklet	7.6		x	[73]
	Aethia pusilla	Least auklet	4.4		x	[74] [72]
	Alca torda	Razorbill	6.2		x	[29]
	Alca torda	Razorbill	6.4		x	[78]
	Cepphus columba	Pigeon guillemot	9.3		x	[29]
	Cepphus columba	Pigeon guillemot	14		x	[42]
	Cepphus columba	Pigeon guillemot	11.8		x	[78]
	Cepphus grylle	Black guillemot	12.1		f	[29]
	Cerorhinca monocerata	Rhinoceros auklet	6.2		x	[29]
	Cerorhinca monocerata	Rhinoceros auklet	7.4		x	[78]
	Fratercula arctica	Atlantic puffin	20		x	[78]
	Fratercula arctica	Atlantic puffin	13.2		x	[29]
	Fratercula cirrhata	Tufted puffin	6		x	[78]
	Ptychoramphus aleuticus	Cassin's auklet	5.7		x	[29]
	Ptychoramphus aleuticus	Cassin's auklet	10.2		x	[78]
	Ptychoramphus aleuticus	Cassin's auklet (+)	6		x	[91]
	Synthliboramphus antiquus	Ancient murrelet	5		x	[29]
	Synthliboramphus antiquus	Ancient murrelet	4.4		x	[51]
	Synthliboramphus hypoleucus	Xantus' murrelet	15		x	[28]
	Synthliboramphus hypoleucus	Xantus' murrelet	9.8		x	[78]
	Uria aalge	Common murre	26.4		x	[29]
	Uria lomvia	Thick-billed murre	22.7		x	[29]
Charadriidae	Charadrius alexandrinus	Snowy plover	7		x	[29]
	Charadrius alexandrinus	Snowy plover (+)	11		x	[78]
	Charadrius alexandrinus	Snowy plover (+)	15		m	[107]
	Charadrius melodus	Piping plover	14		m	[29]
	Charadrius melodus	Piping plover	11		x	[144]
	Charadrius semipalmatus	Semipalmated plover	8.2		x	[29]
	Charadrius vociferus	Killdeer	10.9		x	[29]
	Pluvialis dominica	Lesser golden plover	4.8		x	[78]
	Pluvialis squatarola	Black-bellied plover	12.7		x	[29]
	Pluvialis squatarola	Black-bellied plover	20.3		x	[62]
Haematopodidae	Haematopus bachmani	Black oystercatcher	6.2		x	[29]
	Haematopus bachmani	Black oystercatcher	16		x	[115]
	Haematopus ostralegus	European oystercatcher	36		x	[102]

	Haematopus palliatus	American oystercatcher	14		x	[29]
	Haematopus palliatus	American oystercatcher	17		x	[104]
Laridae	Anous minutus	Black noddy	21.8		x	[78]
	Anous stolidus	Noddy tern	24.9		x	[29]
	Anous tenuirostris	Black noddy	16.1		x	[29]
	Chlidonias niger	Black tern	8.4		x	[29]
	Chlidonias niger	Black tern	17.2		x	[33]
	Gygis alba	Fairy tern	17.8		x	[29]
	Larus argentatus	Herring gull	27.3		x	[29]
	Larus argentatus	Herring gull (+)		44	x	[49]
	Larus argentatus	Herring gull	28		x	[78]
	Larus argentatus	Herring gull	19		x	[90]
	Larus argentatus	Herring gull (+)	30	40	x	[136]
	Larus atricilla	Laughing gull	15.1		x	[29]
	Larus californicus	California gull	13.3		x	[29]
	Larus canus	Mew gull	18.2		x	[29]
	Larus delawarensis	Ring-billed gull	25.1		x	[29]
	Larus delawarensis	Ring-billed gull	31.8		x	[129]
	Larus glaucescens	Glaucous-winged gull	32		x	[25]
	Larus glaucescens	Glaucous-winged gull	22.1		x	[29]
	Larus glaucescens	Glaucous-winged gull	22.8		x	[78]
	Larus marinus	Great black-backed gull	23.3		x	[29]
	Larus occidentalis	Western gull	27.8		x	[29]
	Larus occidentalis	Western gull	25		x	[113]
	Larus pacificus	Pacific gull		14.5	x	[50]
	Larus pipixcan	Franklin's gull (+)	9		x	[29]
	Larus pipixcan	Franklin's gull	9.4		x	[29]
	Pagophila eburnea	Ivory gull	8		x	[78]
	Pagophila eburnea	Ivory gull	17		x	[137]
	Procelsterna cerulea	Blue-gray noddy	11		x	[29]
	Procelsterna cerulea	Blue-gray noddy (+)	14		x	[78]
	Rissa brevirostris	Red-legged kittiwake	4		x	[21]
	Rissa tridactyla	Black-legged kittiwake	5		x	[59]
	Rissa tridactyla	Black-legged kittiwake	7		x	[29]
	Sterna albifrons	Least tern	21.1		x	[29]
	Sterna antillarum	Least tern	24.1		x	[78]
	Sterna caspia	Caspian tern	26.2		x	[29]
	Sterna dougallii	Roseate tern	12.1		x	[29]
	Sterna dougallii	Roseate tern (+)	14		x	[78]
	Sterna elegans	Elegant tern	8.9		x	[29]
	Sterna forsteri	Forster's tern	10.3		x	[29]
	Sterna forsteri	Forster's tern	12		x	[78]
	Sterna fuscata	Sooty tern	32		x	[29]
	Sterna lunata	Gray-backed tern	16.2		x	[29]
	Sterna lunata	Gray-backed tern (+)	18		x	[78]
	Sterna maxima	Royal tern	17		x	[29]
	Sterna maxima	Royal tern	22.2		x	[78]
	Sterna nilotica	Gull-billed tern	6.3		x	[29]
	Sterna nilotica	Gull-billed tern (+)	16		x	[122]
	Sterna paradisaea	Arctic tern	34		x	[29]
	Sterna sandvicensis	Sandwich tern	16.1		x	[29]

Recurvirostridae	Himantopus mexicanus knudsensi	Black-necked (Hawaiian) stilt	5.2		f	[78]
	Recurvirostra americana	American avocet	9		f	[78]
Rynchopidae	Rhynchops niger	Black skimmer	20		x	[30]
Scolopacidae	Arenaria interpres	Ruddy turnstone	14.3		x	[29]
	Bartramia longicauda	Upland sandpiper	5.2		x	[29]
	Calidris alba	Sanderling	5.3		x	[29]
	Calidris alba	Sanderling	13.1		x	[78]
	Calidris alpina	Dunlin	9.3		m	[29]
	Calidris alpina	Dunlin (+)	12.4		x	[78]
	Calidris canutus	Red knot	6.2		x	[29]
	Calidris canutus	Red knot	10.2		x	[78]
	Calidris fuscicollis	White-rumped sandpiper	7		x	[29]
	Calidris himantopus	Stilt sandpiper (+)	11.1		x	[78]
	Calidris mauri	Western sandpiper	9.2		x	[29]
	Calidris melanotos	Pectoral sandpiper	5.9		x	[78]
	Calidris minutilla	Least sandpiper	11.3		x	[29]
	Calidris minutilla	Least sandpiper	16		f	[78]
	Calidris minutilla	Least sandpiper	16		m	[95]
	Calidris ptilocnemis	Rock sandpiper	7.3		x	[78]
	Calidris pusilla	Semipalmated sandpiper	12.1		x	[29]
	Calidris pusilla	Semipalmated sandpiper	12		x	[63]
	Calidris pusilla	Semipalmated sandpiper	12.2		x	[78]
	Catoptrophorus semipalmatus	Willet	8.9		m	[29]
	Gallinago gallinago	Common snipe	9.3		x	[29]
	Limnodromus griseus	Short-billed dowitcher	7.6		x	[29]
	Limnodromus griseus	Short-billed dowitcher	13.3		x	[78]
	Limnodromus scolopaceus	Long-billed dowitcher	8.3		x	[29]
	Limosa fedoa	Marbled godwit	8.6		x	[29]
	Limosa lapponica	Bar-tailed godwit	3.9		x	[78]
	Lobipes lobatus	Northern phalarope	5		f	[29]
	Micropalama himantopus	Stilt sandpiper	9.1		x	[29]
	Numenius arquata	Eurasian curlew	31.5		x	[84]
	Numenius phaeopus	Whimbrel	11		f	[29]
	Numenius phaeopus	Whimbrel		8.8	x	[50]
	Numenius tahitiensis	Bristle-thighed curlew	11.8		x	[78]
	Phalaropus tricolor	Wilson's phalarope	10		m/f	[30]
	Scolopax minor	American woodcock	9.3		f	[29]
	Scolopax minor	American woodcock	11.3		f	[78]
	Tringa flavipes	Lesser yellowlegs	4.8		x	[29]
	Tringa macularia	Spotted sandpiper	12		m	[29]
Stercorariidae	Stercorarius longicaudus	Long-tailed jaeger	8.1		x	[29]
Ciconiiformes						
Ardeidae	Ardea cinerea	Heron		15.8	x	[50]
	Ardea cinerea	Heron		24	x	[138]
	Ardea cocoi	Cocoi heron		24.4	x	[50]
	Ardea goliath	Goliath heron		22.9	x	[50]
	Ardea herodias	Great blue heron	23.3		x	[29]
	Botaurus lentiginosus	American bittern	8.3		x	[29]
	Botaurus lentiginosus	American bittern (+)	8		x	[29]
	Bubulcus ibis	Cattle egret	17		x	[29]

	Bubulcus ibis	Cattle egret	23		x	[135]
	Butorides striatus	Green heron	7.9		x	[29]
	Butorides virescens	Green heron	11.6		x	[29]
	Egretta alba	Great egret	22.8		x	[29]
	Egretta caerulea	Little blue heron	13.9		x	[29]
	Egretta rufescens	Reddish egret	12.3		x	[29]
	Egretta thula	Snowy egret	11.3		x	[29]
	Egretta thula	Snowy egret	17.6		x	[78]
	Hydranassa tricolor	Louisiana heron	17.7		x	[29]
	Nycticorax nycticorax	Black-crowned night heron	21		x	[7] [103] [64]
	Nyticorax nycticorax	Black-crowned night heron	21.1		f	[29]
	Nycticorax violaceus	Yellow-crowned night heron	6.3		x	[78]
Balaenicipitidae	Balaeniceps rex	Shoebill		36	x	[37]
	Balaeniceps rex	Shoebill		35.7	m	[50]
Ciconiidae	Ciconia abdimii	Abdim's stork (+)		21	x	[37]
	Ciconia boyciana	Oriental white stork		48	x	[37]
	Ciconia ciconia	European white stork (+)	33	35	x	[37]
	Ciconia episcopus	Woolly-necked stork (+)		30	x	[37]
	Ciconia maguari	Maguari stork (+)		20	x	[37]
	Ciconia nigra	Black stork	18	31	x	[37]
	Ephippiorhynchus asiaticus	Black-necked stork (+)		34	x	[37]
	Ephippiorhynchus senegalensis	Saddle-billed stork (+)		36	x	[37]
	Jabiru mycteria	Jabiru (+)		36	x	[37]
	Leptoptilos crumeniferus	Marabou	25	41	x	[37]
	Leptoptilos dubius	Greater adjutant		43	x	[37]
	Leptoptilos javanicus	Lesser adjutant		30	x	[37]
	Mycteria americana	Wood stork (+)		27	x	[37]
	Mycteria americana	Wood stork	6.5		x	[78]
	Mycteria ibis	Yellow-billed stork (+)		19	x	[37]
	Mycteria leucocephala	Painted stork (+)		28	x	[37]
Threskiornithidae	Ajaia alaja	Roseate spoonbill	7.8		x	[29]
	Eudocimus albus	White ibis (+)	16.3	20	x	[29] [132]
	Nipponia nippon	Japanese crested ibis		17	x	[37]
	Platalea leucorodia	Eurasian spoonbill	28		x	[37]
	Plegadis chihi	White-faced ibis	14.5	14	x	[29]
	Plegadis falcinellus	Glossy ibis	14.3		x	[29]
	Threskiornis aethiopicus	Sacred ibis	21		x	[37]
	Threskiornis aethiopicus	Sacred ibis		29.2	x	[50]
Columbiformes						
Columbidae	Columba fasciata	Band-tailed pigeon	15.5		x	[30]
	Columba fasciata	Band-tailed pigeon	18.5		x	[78]
	Columba leucocephala	White-crowned pigeon	12.2		x	[30]
	Columba leucocephala	White-crowned pigeon	14.4		x	[78]
	Columba leucocephala	White-crowned pigeon	8.3		x	[90]
	Columba livia	Rock dove		35	x	[48]
	Columba livia	Rock dove (+)	6		x	[71]
	Columbina inca	Inca dove	7.8		x	[30]
	Columbina inca	Inca dove		13.3	x	[50]
	Columbina passerina	Common ground dove	6.1		x	[30]
	Columbina passerina	Common ground dove	7.2		f	[78]

	Geopelia striata	Barred ground dove		14	x	[50]
	Leptotila verreauxi	White-tipped dove	8.6		x	[30]
	Leucosarcia melanoleuca	Wonga-wonga pigeon		16.3	x	[50]
	Streptopelia chinensis	Spotted dove	7.7		x	[30]
	Streptopelia decaocto	Indian turtle dove		17.9	x	[50]
	Zenaida asiatica	White-winged dove	21.8		x	[30]
	Zenaida macroura	Mourning dove	19.3		x	[30][3] [29]
Coraciiformes						
Bucerotidae	Ceratogymna atrata	African black-wattled hornbill		19.3	x	[50]
	Didioceros bicornis	Great Indian hornbill		33.4	x	[50]
Bucorvidae	Bucorvus cafer (leadbeateri)	Southern ground hornbill (+)		70	x	[75]
	Coracias affinis	Burmese roller		8.9	x	[50]
	Coracias gamilus	European roller		8.9	x	[50]
Meropidae	Merops bullockoides	White-fronted bee-eater	7		x	[50]
	Momotus bahamensis	Swainson's motmot		10.4	x	[50]
Cuculiformes						
Coccyzidae	Coccyzus americanus	Yellow-billed cuckoo	5		x	[30]
	Coccyzus erythrophthalmus	Black-billed cuckoo	5		x	[30]
Neomorphidae	Geococcyx californianus	Greater roadrunner (+)	3.8		m	[78]
Dinornithiformes						
Apterygidae	Apteryx australis	Brown kiwi		35	x	[37]
	Apteryx mantelli	Kiwi		20	x	[50]
Falconiformes						
Accipitridae	Accipiter audax	Wedge-tailed eagle		40	x	[37]
	Accipiter cooperii	Cooper's hawk	7.5		f	[29]
	Accipiter cooperii	Cooper's hawk	10.5		f	[78] [101]
	Accipiter cooperii	Cooper's hawk	12		x	[120]
	Accipiter gentilis	Northern goshawk	12.8		f	[29]
	Accipiter gentilis	Northern goshawk		9.4	x	[50]
	Accipiter gentilis	Northern goshawk	13		f	[78]
	Accipiter gentilis	Northern goshawk	19		x	[37]
	Accipiter heliaca	Eastern imperical eagle	55	56	x	[37]
	Accipiter nipalensis	Steppe eagle		41	x	[37]
	Accipiter nisus	Eurasian sparrowhawk	15		x	[37]
	Accipiter striatus	Sharp-shinned hawk	9.8		f	[29] [101]
	Accipiter gentilis	Northern goshawk	19		x	[101]
	Accipiter gentilis	Northern goshawk	13		x	[101]
	Accipiter nisus nisus	Sparrowhawk	15.5		x	[101]
	Aegypius monachus	Eurasian black vulture		39	x	[37]
	Aegypius monachus	Eurasian black vulture	25.5		x	[101]
	Aquila adalberti	Spanish imperial eagle		44.5	x	[50]
	Aquila audax	Wedge-tailed eagle		26.4	x	[50]
	Aquila audax	Wedge-tailed eagle	5.9		x	[101]
	Aquila chrysaetos	Golden eagle	10.4		x	[29]
	Aquila chrysaetos	Golden eagle	17.1		m	[78] [101]
	Aquila chrysaetos	Golden eagle	25.7		x	[101]
	Aquila chrysaetos	Golden eagle	38		x	[37]
	Aquila pomarina	Lesser spotted eagle	26.1		x	[37] [101]
	Aquila pomarina	Lesser spotted eagle		22.3	x	[50]
	Aquila rapax belisarius	Tawny eagle		40.3	f	[50]
	Aquila wahlbergi	Wahlberg's eagle	11.8		x	[101]

Butastur indicus indicus	Gray-faced buzzard	8.5		x	[101]
Buteo buteo	Eurasian buzzard	25		x	[37]
Buteo buteo buteo	Buzzard	25.3		x	[101]
Buteo jamaicensis	Red-tailed hawk	19.8		x	[29]
Buteo jamaicensis	Red-tailed hawk	21.5		x	[78] [101]
Buteo jamaicensis	Red-tailed hawk		29.5	f	[108]
Buteo jamaicensis	Red-tailed hawk	21.4		x	[108]
Buteo lagopus	Rough-legged hawk	17.8		f	[29] [101]
Buteo lagopus	Rough-legged hawk	18		x	[37]
Buteo lagopus	Rough-legged hawk	10.9		x	[101]
Buteo lineatus	Red-shouldered hawk	20		x	[29]
Buteo lineatus	Red-shouldered hawk	19.9		x	[29] [101]
Buteo platypterus	Broad-winged hawk	14.3		x	[29]
Buteo platypterus	Broad-winged hawk	18.3		x	[78] [101]
Buteo regalis	Ferruginous hawk	15.9		x	[29] [101]
Buteo regalis	Ferruginous hawk	20		x	[89] [65]
Buteo swainsoni	Swainson's hawk	15.9		x	[29] [101]
Buteo swainsoni	Swainson's hawk	16		x	[37]
Buteogallus anthracinus	Common black hawk	13.5		x	[124]
Circaetus cinereus	Brown snake eagle	6.8		x	[101]
Circaetus gallicus	Short-toed snake eagle	17		x	[37]
Circaetus gallicus	Short-toed eagle	17.3		x	[101]
Circus aeruginosus	Marsh harrier	16.7		x	[101]
Circus aeruginosus approximans	Swamp harrier	18.2		x	[101]
Circus aeruginosus	Western marsh-harrier	16.4		x	[37]
Circus approximans	Pacific marsh-harrier	18		x	[37]
Circus cyaneus	Northern harrier	16.4		x	[29] [101]
Circus cyaneus	Hen harrier	16		x	[37]
Circus macroarus	Pallid harrier	13.5		x	[101]
Circus pygargus	Montagu's harrier	16		x	[37]
Circus pygargus	Montagu's harrier	16.3		x	[101]
Elanus caeruleus axillanis	Black-shouldered kite	6		x	[101]
Elanus leucurus	White-tailed kite (+)	5.9		x	[29]
Geranoaetus melanoleucus	Chilean eagle		42	x	[50]
Gypaetus barbatus	Bearded vulture (+)		40	x	[37]
Gypaetus barbatus	Bearded vulture		30	x	[50]
Gypohierax angolensis	Sea eagle		27.1	x	[50]
Gyps africanus	White-backed vulture		19.7	x	[50]
Gyps bengalensis	Bengal vulture		17.2	x	[50]
Gyps coprotheres	Cape vulture (griffon)	11.25		x	[101]
Gyps fulvus	Eurasian griffon (+)		41.4	x	[34]
Gyps fulvus	Eurasian griffon		37	x	[37]
Haliaeetus albicilla	White-tailed sea-eagle	27	42	x	[37]
Haliaeetus albicilla	White-tailed eagle	21.1		x	[101]
Haliaeetus leucocephalus	Bald eagle	10.4		x	[29]
Haliaeetus leucocephalus	Bald eagle		47	x	[37]
Haliaeetus leucocephalus	Bald eagle (+)	21.9		m	[78] [101]
Haliaeetus leucocephalus leucocephalus	White-headed sea eagle		33.1	x	[50]
Haliaeetus leucocephalus alascanus	Alaskan bald eagle		28.7	x	[50]

	Haliastur sphenarus	Whistling kite	10.9		x	[101]
	Harpia harpyja	Harpy eagle		16.8	x	[50]
	Harpyopsis novaeguineae	New Guinea eagle		30	x	[37]
	Hieraaetus fasciatus	Bonelli's eagle		20	x	[37]
	Hieraaetus morphnoides	Little eagle	10.3		x	[101]
	Ictinia mississippiensis	Mississippi kite	8		x	[78] [101]
	Kaupifalco monogrammicus	Lizard buzzard	8.6		x	[101]
	Lophoictinia isura	Square-tailed kite (+)	17		x	[2]
	Milvus migrans	Black kite	23.7		x	[101]
	Milvus migrans	Black kite	23		x	[37]
	Milvus milvus	Red kite		38	x	[37]
	Milvus milvus	Red kite	26		x	[84]
	Milvus milvus	Red kite	25.8		x	[101]
	Neophron percnopterus	Egyptian vulture		37	x	[37]
	Parabuteo unicinctus	Harris' hawk		25	f	[4]
	Parabuteo unicinctus	Harris' hawk	11		x	[29]
	Parabuteo unicinctus	Harris' hawk	10		x	[29]
	Parabuteo unicinctus	Harris' hawk	12.6		f	[78] [101]
	Pernis apivorus	Western honey-buzzard	29		x	[37] [101]
	Rostrhamus sociabilis	Snail kite	17		x	[6]
	Rostrhamus sociabilis	Snail kite	7		x	[78] [101]
Cathartidae	Cathartes aura	Turkey vulture	16.8		x	[29] [101]
	Coragyps atratus	Black vulture	25.5		x	[29]
	Gymnogyps californianus	California condor	45		x	[37]
	Gymnogyps californianus	California condor		36.5	x	[50]
	Sarcorhamphus papa	King vulture		40	x	[46]
	Sarcorhamphus papa	King vulture		23.3	x	[50]
Falconidae	Falco berigora	Brown falcon	10.5		x	[101]
	Falco berigora	Brown falcon		16	x	[37]
	Falco columbarius	Merlin	7.8		x	[29] [101]
	Falco columbarius	Merlin	12.5		x	[101]
	Falco columbarius	Merlin	8		x	[106]
	Falco eleonorae	Eleonora's falcon	6		x	[101]
	Falco mexicanus	Prairie falcon	9.1		m	[29]
	Falco mexicanus	Prairie falcon	10.6		f	[78] [101]
	Falco naumanni	Lesser kestrel	6.2		x	[101]
	Falco peregrinus	Peregrine falcon	12.3		x	[29] [101]
	Falco peregrinus	Peregrine falcon	16.3		x	[101]
	Falco rupicoloides	White-eyed kestrel	6.5		x	[101]
	Falco sparverius	American kestrel	11.6		f	[29] [101]
	Falco subbuteo longipennis	Little falcon	10.5		x	[101]
	Falco subbuteo subbuteo	European hobby	10.8		x	[101]
	Falco tinnunculus	Common kestrel	16.2		x	[101]
	Falco tinnunculus	Common kestrel	16		x	[37]
	Falco vespertinus vespertinus	Red-footed falcon	12.3		x	[101]
	Polyborus cheriway	Cheriway caracara		25.8	x	[50]
	Polyborus plancus	Crested caracara	8.9		x	[78]
	Polyborus tharus	Brazilian caracara		37.6	x	[50]
Pandionidae	Pandion haliaetus	Osprey	21.9		x	[29]
	Pandion haliaetus	Osprey	25		x	[37] [101]

	Pandion haliaetus	Osprey	23		x	[78]
	Pandion haliaetus	Osprey	24.8		x	[101]
Sagittariidae	Sagittarius serpentarius	Secretary bird		18.6	x	[50] [5]
Galliformes						
Cracidae	Crax globulosa	Wattled curassow (+)		20	x	[37]
	Crax rubra	Great curassow		24	x	[37]
	Mitu tuberosa	Razor-billed curassow		23	x	[37]
	Ortalis vetula	Plain chachalaca	0.2		m	[29]
Megapodiidae	Macrocephalon maleo			23	x	[37]
Odontophoridae	Callipepla gambelii	Gambel's quail	7.4		m	[29]
	Callipepla californica	California quail	6.9		x	[29]
	Colinus virginianus	Common bobwhite	6.4		m	[29]
Phasianidae	Alectoris rufa	Red-legged partridge		6.25	x	[50]
	Bonasa umbellus	Ruffed grouse	7.6		x	[29]
	Bonasa umbellus	Ruffed grouse	8.5		f	[78]
	Dendragapus canadensis	Spruce grouse	13		f	[10]
	Dendragapus canadensis	Spruce grouse	13		m	[10]
	Dendragapus canadensis	Spruce grouse	5.3		f	[29]
	Dendragapus obscurus	Blue grouse	10.9		m	[29]
	Dendragapus obscurus	Blue grouse (+)	14		m	[147]
	Dendragapus obscurus	Blue grouse (+)	11		f	[147]
	Gallus gallus	Red junglefowl		30	x	[98]
	Lagopus leucurus	White-tailed ptarmigan	15		m	[13]
	Lagopus leucurus	White-tailed ptarmigan	12		f	[13]
	Meleagris gallopavo	Wild turkey	12.5		x	[29]
	Meleagris gallopavo	Wild turkey		12.3	x	[50]
	Meleagris gallopavo	Wild turkey	13		f	[145]
	Pavo cristatus	Peafowl		23.2	x	[50]
	Perdix perdix	Gray partridge	5.17		x	[27]
	Perdix perdix	Gray partridge (+)		5	x	[50]
	Phasianus colchicus	Common pheasant		27	x	[50]
	Tetrao urogallus	Western capercaillie		18	x	[37]
	Tympanuchus cupido	Greater prairie chicken	1.5		x	[119]
	Tympanuchus pallidicinctus	Lesser prairie chicken	13.5		x	[29]
	Tympanuchus phasianellus	Sharp-tailed grouse	6.3		m	[29]
Gaviiformes						
Gaviidae	Gavia arctica	Black-throated diver	28		x	[37]
	Gavia immer	Common loon	7.8		x	[29]
	Gavia immer	Great northern diver(+)	20		x	[37]
	Gavia stellata	Red-throated diver	23		x	[37]
Gruiformes						
Gruidae	Grus americana	Whooping crane	30		x	[8] [85] [97]
	Grus americana	Whooping crane	7.3		x	[78]
	Grus americana	Whooping crane		40	x	[100]
	Grus antigone	Sarus crane		26	x	[50]
	Grus canadensis	Sandhill crane	18.3		x	[29]
	Grus canadensis	Sandhill crane	19.4		x	[78]
	Grus canadensis	Sandhill crane	21.5		x	[134]
	Grus grus lilfordi	Lord Lilford's crane		39.4	x	[50]
	Grus japonensis	Manchurian crane		25.2	x	[50]
	Grus leucauchen	White-necked crane		25.8	x	[50]

	Grus leucogeranus	Asiatic white crane		36.2	m	[50]
	Grus leucogeranus	Asiatic white crane		32.3	f	[50]
	Grus monachus	Hooded crane		24	x	[50]
	Grus rubicunda			33	x	[50]
Otididae	Otis tarda	Great bustard		28.3	x	[50]
Rallidae	Fulica americana	American coot	19.5		x	[29]
	Fulica americana	American coot	22.3		f	[78]
	Gallinula chloropus	Common gallinule	10.5		x	[29]
	Gallirallus australus	Weka		13.8	x	[50]
	Gallirallus brachypterus	New Zealand weka rail		12.3	x	[50]
	Laterallus jamicensis	Black rail	2.4		m	[47]
	Rallus longirostris	Clapper rail	7.5		m	[29]
Musophagiformes						
Musophagidae	Turacus leucotis donaldsoni	Donaldson Smith's touraco		12.4	x	[50] [77]
Passeriformes						
Aegithalidae	Psaltiparus minimus	Bushtit (+)	8.4		m	[77]
Alaudidae	Eremophila alpestris	Horned lark	7.1		m	[30]
	Eremophila alpestris	Horned lark	7.9		x	[78]
	Eremophila alpestris	Horned lark	8		x	[78]
Bombycillidae	Bombycilla cedrorum	Cedar waxwing	7		x	[77]
	Bombycilla garrulus	Bohemian waxwing	5.8		x	[77]
Certhiidae	Certhia familiaris	Brown creeper (+)	4.6		x	[77]
Cinclidae	Cinclus mexicanus	American dipper (+)	5.1		m	[77]
Corcoracidae	Corcorax melanorhamphos	White-winged chough		16.3	x	[50]
	Struthidea cinerea	Apostlebird		25	x	[50]
Corvidae	Aphelocoma coerulescens	Scrub jay	15.8		x	[30]
	Aphelocoma ultramarina	Mexican jay	19		m	[16]
	Aphelocoma ultramarina	Gray-breasted (Mexican) jay	10.8		x	[30]
	Aphelocoma ultramarina	Gray-breasted jay	17.6		x	[78]
	Corvus brachyrhynchos	American crow	14.6		x	[30]
	Corvus brachyrhynchos	American crow	13.8		x	[90]
	Corvus caurinus	Northwestern crow	12.8		x	[30]
	Corvus corax	Common raven	13.3		x	[30]
	Corvus corax	Raven		24.7	x	[50]
	Corvus cornix	Hooded crow		16.8	x	[50]
	Corvus corone	Carrion crow		13	f	[50]
	Corvus corone	Carrion crow		11	m	[50]
	Corvus coronoides	Australian crow or raven		21.8	x	[50]
	Corvus cryptoleucus	Chihuahuan (white-necked) raven	12.5		x	[30]
	Corvus frugilegus	Rook		14	f	[50]
	Corvus ossifragus	Fish crow	14.5		x	[30]
	Cyanocitta cristata	Blue jay	18.3		x	[30]
	Cyanocitta cristata	Blue jay	16.3		x	[78]
	Cyanocitta stelleri	Steller's jay	11		m	[30]
	Cyanocitta stelleri	Steller's jay	16.1		x	[78]
	Cyanocorax caeruleus	Azure jay (+)		36	x	[31]
	Cyanocorax yncas	Green jay	10.7		x	[30]
	Garrulus glandavius	Jay		17	x	[50] [112]
	Gymnorhinus cyanocephalus	Pinyon jay	8.8		f	[30]
	Gymnorhinus cyanocephalus	Pinyon jay	11		x	[78]

	Nucifraga columbiana	Clark's nutcracker	17.4		x	[30]
	Perisoreus canadensis	Gray jay	19.2		x	[30]
	Perisoreus canadensis	Gray jay	16		f	[131]
	Perisoreus canadensis	Gray jay (+)	14		m	[131]
	Pica nuttalli	Yellow-billed magpie	9.9		f	[30]
	Pica nuttalli	Yellow-billed magpie (+)	10		x	[117]
	Pica pica	Black-billed magpie	5		x	[30]
	Pica pica	Black-billed magpie	5.9		x	[78]
	Pyrrhocorax pyrrhocorax	Cornish chough		20.3	x	[50]
Cotingidae	Rupicola rupicola	Cock-of-the-rock		7	x	[50]
Drepanididae	Hemignathus virens	Common amakihi	7.7		x	[78]
	Himatione sanguinea	Apapane	6.4		x	[79]
	Telespyza cantans	Laysan finch (+)	6		m	[79]
	Telespyza ultima	Nihoa finch (+)	11		x	[79]
Emberizidae	Aimophila aestivalis	Bachman's sparrow	3		x	[57]
	Aimophila quinquestriata	Five-striped sparrow	4		x	[96]
	Aimophila ruficeps	Rufous-crowned sparrow (+)	3.2		x	[79]
	Ammodramus caudacutus	Sharp-tailed sparrow	10		m	[54]
	Ammodramus caudacutus	Sharp-tailed sparrow	6		f	[54]
	Ammodramus maritimus	Seaside sparrow	8.9		x	[78]
	Ammodramus maritimus	Seaside sparrow	9		m	[133] [114]
	Ammodramus maritimus	Seaside sparrow	9		f	[133] [114]
	Ammodramus maritimus nigrescens	(Dusky) seaside sparrow	8.3		m	[79]
	Ammodramus savannarum	Grasshopper sparrow (+)	3.1		x	[79]
	Amphispiza bilineata	Black-throated sparrow (+)	6		f	[79]
	Calamospiza melanocorys	Lark bunting (+)	4		m	[79]
	Calcarius lapponicus	Lapland longspur (+)	5		x	[79]
	Calcarius pictus	Smith's longspur	5		m	[14]
	Calcarius pictus	Smith's longspur (+)	4		f	[14]
	Cardinalis cardinalis	Northern cardinal	15.8		f	[79]
	Cardinalis sinuatus	Pyrrhuloxia (+)	5.8		m	[79]
	Chondestes grammacus	Lark sparrow (+)	7.8		x	[79]
	Cyanerpes caeruleus	Purple sugar bird		17	x	[50]
	Guiraca caerulea	Blue grosbeak	5.9		m	[79]
	Junco hyemalis ssp.	Dark-eyed (slate-colored) junco	10.8		m	[79]
	Junco hyemalis ssp.	Dark-eyed (gray-headed) junco (+)	10.7		x	[79]
	Junco hyemalis ssp.	Dark-eyed (Oregon) junco (+)	9.8		x	[79]
	Junco hyemalis ssp.	Dark-eyed (white-winged) junco (+)	6.6		x	[79]
	Junco phaeonotus ssp.	Yellow-eyed (Mexican junco (+)	6.6		x	[79]
	Melospiza georgiana	Swamp sparrow (+)	4.8		x	[79]
	Melospiza lincolnii	Lincoln's sparrow	7.6		x	[79]
	Melospiza melodia	Song sparrow (+)	11.3		x	[79]
	Passerculus sandwichensis	Savannah sparrow (+)	6.1		x	[79]
	Passerella iliaca	Fox sparrow (+)	9.7		x	[79]
	Passerina amoena	Lazuli bunting	8.9		x	[78]
	Passerina amoena	Lazuli bunting	4.9		m	[79]

	Passerina ciris	Painted bunting (+)	10.5		f	[79]
	Passerina cyanea	Indigo bunting	11		m	[87] [110]
	Passerina cyanea	Indigo bunting	11		f	[87] [110]
	Pheucticus ludovicianus	Rose-breasted grosbeak	7.3		x	[78]
	Pheucticus melanocephalus	Black-headed grosbeak		25	x	[143]
	Pheucticus melanocephalus	Black-headed grosbeak	9.1		m	[79]
	Pipilo aberti	Abert's towhee	8.6		x	[79]
	Pipilo chlorurus	Green-tailed towhee (+)	7.7		X	[79]
	Pipilo erythrophthalmus	Rufous-sided towhee (+)	12.3		m	[79]
	Pipilo erythrophthalmus ssp.	Rufous-sided (spotted) towhee (+)	10.7		m	[79]
	Pipilo fuscus	Brown towhee (+)	12.8		m	[79]
	Piranga ludoviciana	Western tanager (+)	7.9		m	[79]
	Piranga olivicea	Scarlet tanager (+)	10.1		m	[79]
	Piranga rubra	Summer tanager	7.3		x	[78]
	Piranga rubra	Summer tanager (+)	6		m	[79]
	Plectrophenax hyperboreus	Mckay's bunting	4.8		f	[79]
	Plectrophenax nivalis	Snow bunting	8.8		x	[79]
	Pooecetes gramineus	Vesper sparrow	7.9		x	[78]
	Pooecetes gramineus	Vesper sparrow (+)	6.9		x	[79]
	Spiza americana	Dickcissel	4		m	[79]
	Spizella arborea	American tree sparrow	10.1		x	[79]
	Spizella breweri	Brewer's sparrow	5.2		x	[79]
	Spizella pallida	Clay-colored sparrow	5.1		x	[79]
	Spizella passerina	Chipping sparrow	9.8		x	[79]
	Spizella pusilla	Field sparrow	8.8		x	[78]
	Spizella pusilla	Field sparrow (+)	7.8		x	[79]
	Sporophila torqueola	White-collared seedeater	11.7		f	[79]
	Zonotrichia albicollis	White-throated sparrow	9.7		x	[79]
	Zonotrichia atricapilla	Golden-crowned sparrow (+)	10.5		x	[79]
	Zonotrichia leucophrys ssp.	(Gambel's) white-crowned sparrow	13.3		x	[79]
	Zonotrichia querula	Harris' sparrow	11.7		x	[79]
Fringillidae	Carduelis flammea	Common redpoll (+)	5.8		m	[79]
	Carduelis hornemanni	Hoary redpoll	4.6		x	[78]
	Carduelis pinus	Pine siskin	9.4		x	[79]
	Carduelis psaltria	Lesser goldfinch	5.7		f	[79]
	Carduelis tristis	American goldfinch	11		x	[94]
	Carduelis tristis	American goldfinch	9.3		x	[78]
	Carpodacus cassinii	Cassin's finch (+)	7		m	[79]
	Carpodacus mexicanus	House finch	1.1		x	[78]
	Carpodacus mexicanus	House finch	11.6		m	[79]
	Carpodacus purpureus	Purple finch (+)	11.8		m	[79]
	Coccothraustes vespertinus	Evening grosbeak	15.3		m	[79]
	Fringilla coelebs	Chaffinch		29		[99]
	Leucosticte arctoa atrata	(Black) rosy finch	5.7		x	[79]
	Leucosticte arctoa ssp.	(Gray-crowned) rosy finch(+)	6.6		f	[79]
	Loxia curvirostra	Red crossbill	4.2		x	[79]
	Loxia leucoptera	White-winged crossbill		4	x	[2]
	Pinicola enucleator	Pine grosbeak	9.7		m	[79]
	Serinus canarius	House canary		24	x	[82]

	Serinus canarius domesticus	House canary		22	m	[50]
Hirundinidae	Hirundo fulva	Cave swallow	4.8		x	[78]
	Hirundo fulva	Cave swallow	9.3		x	[142]
	Hirundo pyrrhonota	Cliff swallow	11		x	[15]
	Hirundo pyrrhonota	Cliff swallow	9		m	[30]
	Hirundo pyrrhonota	Cliff swallow	10		x	[78]
	Hirundo rustica	Barn swallow	8.1		f	[30]
	Progne subis	Purple martin	13.8		x	[30]
	Riparia riparia	Bank swallow	8		f	[30]
	Stelgidopteryx ruficollis	Rough-winged swallow	5.9		x	[30]
	Tachycineta bicolor	Tree swallow	11		f	[30]
	Tachycineta bicolor	Tree swallow	8.9		x	[68] [66] [20]
	Tachycineta thalassina	Violet-green swallow	6.8		m	[30]
Icteridae	Agelaius phoeniceus	Red-winged blackbird	15.8		m	[79]
	Dolichonyx oryzivorus	Bobolink	9		x	[11]
	Dolichonyx oryzivorus	Bobolink	8.1		x	[78]
	Euphagus carolinus	Rusty blackbird	8.8		x	[32]
	Euphagus cyanocephalus	Brewer's blackbird (+)	11		m	[79]
	Icterus cucullatus	Hooded oriole	6		m	[79]
	Icterus galbula ssp.	Northern (Baltimore) oriole	11.6		f	[79]
	Icterus galbula ssp.	Northern (bullock's) oriole	6.6		m	[79]
	Icterus parisorum	Scott's oriole (+)	6.4		m	[79]
	Icterus pectoralis	Spot-breasted oriole (+)	11.6		x	[79]
	Icterus spurius	Orchard oriole	9.3		x	[78]
	Icterus spurius	Orchard oriole	9.6		x	[79]
	Molothrus aeneus	Bronzed cowbird	3.6		m	[79]
	Molothrus ater	Brown-headed cowbird	16.9		x	[78]
	Molothrus ater	Brown-headed cowbird	15.8		m	[79]
	Quiscalus major	Boat-tailed grackle (+)	11.9		x	[79]
	Quiscalus mexicanus	Great-tailed grackle	12.5		m	[79]
	Quiscalus quiscula	Common grackle	20.9		m	[79]
	Sturnella magna	Eastern meadowlark	9		x	[86]
	Sturnella neglecta	Western meadowlark		10	m	[86]
	Sturnella neglecta	Western meadowlark		9	f	[86]
	Xanthocephalus xanthocephalus	Yellow-headed blackbird		16	m	[1]
	Xanthocephalus xanthocephalus	Yellow-headed blackbird	10.9		m	[79]
Irenidae	Irena turcosa	Turquoise fairy bluebird (+)		16.1	x	[50]
Laniidae	Lanius excubitor	Northern shrike (+)	3.3		x	[77]
	Lanius ludovicianus	Loggerhead shrike	6.1		x	[78]
	Lanius ludovicianus	Loggerhead shrike (+)	12.5		x	[77]
Mimidae	Dumetella carolinensis	Gray catbird	10.9		x	[77]
	Mimus polyglottos	Northern mockingbird	20		x	[61]
	Toxostoma bendirei	Bendire's thrasher	4.2		x	[77]
	Toxostoma curvirostre	Curve-billed thrasher	10.8		x	[77]
	Toxostoma dorsale	Crissal thrasher	4.4		x	[77]
	Toxostoma lecontei	Leconte's thrasher	5.7		m	[77]
	Toxostoma longirostre	Long-billed thrasher	7.3		x	[77]
	Toxostoma redivivum	California thrasher	6.9		x	[77]
	Toxostoma rufum	Brown thrasher (+)	12.8		x	[77]
Motacillidae	Anthus rubescens	American pipit	5.1		x	[77]

	Anthus spinoletta	Water pipit (+)	5.1		f	[77]
Muscicapidae	Muscicapa striata	Spotted flycatcher		11.8	x	[50]
Nectariniidae	Nectarinia asiatica	Purple sunbird		22	m	[50]
Oriolidae	Orioles xanthornus	Black-hooded oriole (hang-nest)		15.5	x	[50]
	Sphecotheres viridis	Green figbird (hang-nest)		18	f	[50]
Paradisaeidae	Paradisea apoda	Bird of paradise		12.1	x	[50]
Paridae	Parus atricapillus	Black-capped chickadee	12.4		x	[30] [128]
	Parus bicolor ssp.	Tufted (eastern) titmouse	13.3		x	[30]
	Parus bicolor ssp.	Tufted (western) titmouse	4.8		x	[30]
	Parus carolinensis	Carolina chickadee	10.9		x	[30]
	Parus cinctus	Siberian tit	7		x	[139]
	Parus gambeli	Moutain chickadee	10.1		m	[30]
	Parus hudsonicus	Boreal chickadee	4.7		x	[78]
	Parus inornatus	Plain titmouse	8		x	[30]
	Parus major	Great titmouse		9	x	[50]
	Parus rufescens	Chestnut-backed chickadee	7.8		x	[30]
	Parus sclateri	Mexican chickadee	5		x	[44]
	Parus wollweberi	Bridled titmouse	6.6		x	[30]
Parulidae	Cardellina rubrifrons	Red-faced warbler (+)	3		m/f	[92]
	Dendroica caerulescens	Black-throated blue warbler	9.7		x	[78]
	Dendroica caerulescens	Black-throated blue warbler (+)	8		f	[77]
	Dendroica castanea	Bay-breasted warbler	3.4		x	[78]
	Dendroica coronata sp.	Yellow-rumped (Audubon's) warbler	6.6		x	[77]
	Dendroica coronata sp.	Yellow-rumped (myrtle) warbler	6.9		x	[77]
	Dendroica discolor	Prairie warbler (+)	10.3		m	[77]
	Dendroica dominica	Yellow-throated warbler	5.1		x	[78]
	Dendroica fusca	Blackburnian warbler	8.2		m	[78]
	Dendroica kirtlandii	Kirtland's warbler	9		m	[93][140]
	Dendroica kirtlandii	Kirtland's warbler	8		f	[93][140]
	Dendroica magnolia	Magnolia warbler	6.9		x	[111]
	Dendroica palmarum hypochrysea	Yellow palm warbler (+)	5.8		x	[77]
	Dendroica palmarum palmarum	Western palm warbler (+)	6.6		x	[77]
	Dendroica pensylvanica	Chestnut-sided warbler	6.9		x	[77]
	Dendroica petechia	Yellow warbler (+)	8.9		m	[77]
	Dendroica pinus	Pine warbler (+)	6.8		x	[77]
	Dendroica striata	Blackpoll warbler	4.3		x	[78]
	Dendroica striata	Blackpoll warbler	3.4		f	[77]
	Dendroica tigrina	Cape May warbler (+)	4.3		f	[77]
	Dendroica townsendi	Townsend's warbler	3.6		x	[78]
	Dendroica virens	Black-throated green warbler	5.9		m	[77]
	Geothlypis trichas	Common yellowthroat (+)	9.9		f	[77]
	Helmitheros vermivorus	Worm-eating warbler (+)	7.1		x	[77]
	Icteria virens	Yellow-breasted chat (+)	8.9		x	[77]
	Limnothlypis swainsonii	Swainson's warbler	7.8		x	[77]
	Mniotilta varia	Black and white warbler	11.3		f	[77]

	Myioborus pictus	Painted redstart	6.6		x	[78]
	Myioborus pictus	Painted redstart	4.3		m	[77]
	Oporonis tolmiei	Macgillivray's warbler	4.1		x	[78]
	Oporornis agilis	Connecticut warbler	4.3		x	[77]
	Oporornis formosus	Kentucky warbler (+)	6		m	[77]
	Oporornis philadelphia	Mourning warbler (+)	7.9		m	[77]
	Oporornis tolmiei	Macgillivray's warbler	4.1		m	[78]
	Parula americana	Northern parula (+)	4.9		f	[77]
	Protonotaria citrea	Prothonotary warbler	4.9		m	[77]
	Seiurus aurocapillus	Ovenbird	9		x	[77]
	Seiurus motacilla	Louisiana waterthrush	8		x	[53]
	Seiurus novaeboracensis	Northern waterthrush	8.9		x	[78]
	Setophaga ruticilla	American redstart	10		m	[77]
	Vermivora celata	Orange-crowned warbler (+)	6.8		m	[77]
	Vermivora chrysoptera	Golden-winged warbler (+)	7.9		m	[77]
	Vermivora leucobronchialis	Brewster's warbler (+)	6		m	[77]
	Vermivora luciae	Lucy's warbler	5.8		x	[78]
	Vermivora peregrina	Tennessee warbler (+)	6.4		f	[77]
	Vermivora pinus	Blue-winged warbler	7.9		x	[77]
	Vermivora ruficapilla	Nashville warbler	7.3		f	[77]
	Vermivora virginiae	Virginia's warbler	4.1		x	[78]
	Vermivora virginiae	Virginia's warbler	3.2		x	[77]
	Wilsonia canadensis	Canada warbler	8		x	[78]
	Wilsonia canadensis	Canada warbler	7.9		f	[77]
	Wilsonia citrina	Hooded warbler	8		x	[77]
	Wilsonia pusilla	Wilson's warbler (+)	6.8		m	[77]
Passeridae	Passer domesticus	House sparrow	13.3		x	[79]
	Passer italiae	Italian sparrow		20	x	[98]
	Passer montanus	Eurasian tree sparrow	4		x	[79]
	Bubalornis albirostris	African ox birds		20	x	[50]
Polioptilidae	Auriparis flaviceps	Verdin (+)	5.6		x	[77]
	Polioptila caerulea	Blue-gray gnatcatcher (+)	4.2		x	[77]
Prunellidae	Prunella modularis occidentalus	Hedge sparrow		8.3	x	[50]
Pycnonotidae	Pycnonotus jocusus	Red-whiskered bulbul	4		x	[77]
Regulidae	Regulus calendula	Ruby-crowned kinglet (+)	5.6		x	[77]
	Regulus satrapa	Golden-crowned kinglet (+)	5.3		m	[77]
Sittidae	Sitta canadensis	Red-breasted nuthatch	7.5		m	[77]
	Sitta carolinensis	White-breasted nuthatch	9.8		x	[77]
	Sitta pusilla	Brown-headed nuthatch (+)	5.8		x	[77]
	Sitta pygmaea	Pygmy nuthatch (+)	8.2		x	[77]
Sturnidae	Acridotheres cristatellus	Crested myna		13	x	[50]
	Acridotheres cristatellus	Crested myna	11.2		x	[70]
	Lamprotornis caudatus	Long-tailed glossy starling		17.8	x	[50]
	Spreo hildebrandti	Hildebrandt's starling		16.9	x	[50]
	Spreo hildebrandti	Hildebrandt's starling		16	f	[50]
	Sturnus vulgaris	European starling	21.3		x	[43]
	Sturnus vulgaris	European starling (+)		15.8		[98] [50]
Sylviidae	Acrocephalus familiaris	Millerbird (+)	4.2		x	[77]
	Acrocephalus stentorous brunnescens	Indian great reed warbler		6.6	x	[50]
	Chamaea fasciata	Wrentit (+)	12.6		x	[77]

	Phylloscopus borealis	Arctic warbler	3.1		m	[77]
	Sylvia borin	Garden warbler		24	m	[50]
	Sylvia communis communis	Whitethroat		6.3	x	[50]
Timaliidae	Leiothrix lutea	Red-billed leothrix		15	x	[50]
	Turdoides temicolor	Bengal jungle babbler		16.5	x	[50]
Troglodytidae	Campylorhynchus brunneicapillus	Cactus wren	6.3		x	[77]
	Thryomanes bewickii	Bewick's wren	8		x	[78]
	Thryomanes bewickii	Bewick's wren (+)	7.6		x	[77]
	Thryothorus ludovicianus	Carolina wren	6.1		x	[77]
	Troglodytes aedon	House wren (+)	7.1		x	[77]
	Troglodytes troglodytes	Winter wren (+)	4.1		m	[77]
Turdidae	Catharus fuscescens	Veery	10.1		x	[39]
	Catharus guttatus	Hermit thrush (+)	8.7		x	[77]
	Catharus minimus	Gray-cheeked thrush (+)	7.3		x	[77]
	Catharus ustulatus	Swainson's thrush (+)	10.1		f	[77]
	Hylocichla mustelina	Wood thrush (+)	8.9		x	[77]
	Ixoreus naevius	Varied thrush	4.8		x	[78]
	Ixoreus naevius	Varied thrush (+)	3.6		m	[77]
	Sialia currucoides	Mountain bluebird	4.8		x	[77]
	Sialia mexicanus	Western bluebird (+)	4.3		m	[77]
	Sialia sialis	Eastern bluebird	8		x	[77]
	Turdus migratorius	American robin (+)	13.9		x	[77]
	Turdus migratorius	American robin		12.8	x	[98]
Tyrannidae	Camptostoma imberbe	Northern beardless tyrannulet	4.8		x	[78]
	Contopus borealis	Olive-sided flycatcher	5.9		x	[30]
	Contopus borealis	Olive-sided flycatcher	7.1		x	[78]
	Contopus sordidulus	Western wood pewee	6.1		x	[30]
	Contopus virens	Eastern wood pewee	7.1		m	[30]
	Empidonax difficilis	Western flycatcher	6		f	[30]
	Empidonax flaviventris	Yellow-bellied flycatcher	3.9		x	[30]
	Empidonax fulvifrons	Buff-breasted flycatcher (+)	2		x	[12]
	Empidonax minimus	Least flycatcher	5.9		x	[30]
	Empidonax minimus	Least flycatcher	8		x	[78]
	Empidonax oberholseri	Dusky flycatcher	9		f	[30]
	Empidonax oberholseri	Dusky flycatcher	8.2		x	[78]
	Empidonax oberholseri	Dusky flycatcher (+)	8		f	[127]
	Empidonax oberholseri	Dusty flycatcher (+)	6		x	[127]
	Empidonax traillii	Willow (Traill's) flycatcher	7		x	[30]
	Empidonax virescens	Acadian flycatcher	10.9		m	[30]
	Myiarchus cinerascens	Ash-throated flycatcher	4.9		x	[30]
	Myiarchus cinerascens	Ash-throated flycatcher	9		x	[78]
	Myiarchus crinitus	Great-crested flycatcher	13.9		x	[30]
	Myiarchus tyrannulus	Brown-crested (Wied's) flycatcher	8		x	[30]
	Myiarchus tyrannulus	Brown-crested flycatcher	9.9		x	[78]
	Pyrocephalus rubinus	Vermilion flycatcher	5.5		x	[78]
	Sayornis nigricans	Black phoebe	3		x	[30]
	Sayornis nigricans	Black phoebe	8		x	[78]
	Sayornis phoebe	Eastern phoebe	9.1		x	[30]
	Tyrannus tyrannus	Eastern kingbird	9.9		x	[30]

	Tyrannus verticalis	Western kingbird	6.9		x	[30]
Vireonidae	Vireo atricapillus	Black-capped vireo	7		x	[55]
	Vireo bellii	Bell's vireo (+)	6.9		x	[77]
	Vireo flavifrons	Yellow-throated vireo (+)	6.1		x	[77]
	Vireo gilvus	Warbling vireo (+)	13.1		m	[77]
	Vireo griseus	White-eyed vireo (+)	7.3		x	[77]
	Vireo huttoni	Hutton's vireo	6.5		f	[36] [78]
	Vireo olivaceus	Red-eyed vireo (+)	10		x	[77]
	Vireo philadelphicus	Philadelphia vireo	8.8		x	[77]
	Vireo solitarius	Solitary vireo	7.4		x	[77]
Zosteropidae	Zosterops japonicus	Japanese white-eye	5.1		x	[78]
Pelecaniformes						
Anhingidae	Anhinga anhinga	American anhinga	11.9		x	[29]
	Anhinga anhinga	American anhinga		16.4	x	[91]
	Anhinga melanogaster	Darter (+)		16	x	[37]
Fregatidae	Fregata ariel	Lesser frigatebird	17		m	[29]
	Fregata ariel	Lesser frigatebird	23.5		x	[78]
	Fregata magnificens	Magnificent frigatebird	7.2		f	[29]
	Fregata minor	Great frigatebird	30		f	[29]
	Fregata minor	Great frigatebird	34		x	[37]
Pelecanidae	Pelecanus crispus	Dalmation pelican		35.3	x	[50]
	Pelecanus erythrorhynchos	American white pelican		54	x	[29]
	Pelecanus erythrorhynchos	American white pelican	16		x	[37]
	Pelecanus erythrorhynchos	American white pelican	26.4		x	[29]
	Pelecanus occidentalis	Brown pelican		31	x	[37]
	Pelecanus onocrotalus	White pelican		51		[50]
	Pelecanus occidentalis	Brown pelican	19.7		x	[29]
Phaethontidae	Phaethon aethereus	Red-billed tropicbird	5.25		x	[29]
	Phaethon aethereus	Red-billed tropicbird (+)	10		x	[37]
	Phaethon rubricauda	Red-tailed tropicbird	16.1		x	[29]
	Phaethon rubricauda	Red-tailed tropicbird	9		x	[37]
	Phaethon rubricauda	Red-tailed tropicbird	23.8		x	[78]
Phalacrocoracidae	Phalacrocorax aristotelis	Shag		14.3	x	[50]
	Phalacrocorax auritas	Double-breasted cormorant	17.8		x	[78] [29]
	Phalacrocorax carbo	Great cormorant	14.3		x	[29]
	Phalacrocorax carbo	Great cormorant	18.0		x	[37]
	Phalacrocorax olivaceus	Olivaceous cormorant	12.6		x	[29]
	Phalacrocorax pelagicus	Pelagic cormorant	17.8		x	[29]
	Phalacrocorax penicillatus	Brandt's cormorant	9.4		x	[29]
	Phalacrocorax purpurascens	Macquarie shag	13.0		x	[37]
Sulidae	Morus bassanus	Northern gannet	20.3		x	[29]
	Morus bassanus	Gannet		12.8	x	[50]
	Sula (= Morus) bassanus	Northern gannet (+)	25.0		x	[37]
	Sula (= Morus) bassanus	Northern gannet	20.4		x	[78]
	Sula dactylatra	Masked booby	25.3		x	[29]
	Sula dactylatra	Masked booby (+)	23.0		x	[37]
	Sula leucogaster	Brown booby	16.1		f	[29]
	Sula leucogaster	Brown booby	24.6		f	[78]
	Sula serrator	Australian gannet (+)	33.0		x	[37]
	Sula sula	Red-footed booby	21.9		x	[29]
	Sula sula	Red-footed booby	23.0		x	[37]

Phoenicopteriformes						
Phoenicopteridae	Phoenicopterus ruber	American flamingo	13.2		x	[29]
	Phoenicopterus ruber	Greater flamingo (+)	33.0	44.0	x	[37]
Piciformes						
Picidae	Colaptes auratus	Northern flicker	12.5		x	[30]
	Colaptes auratus ssp.	Northern (yellow-shafted) flicker	9.2		m	[30]
	Colaptes auratus ssp.	Northern (red-shafted) flicker	6.7		x	[30]
	Colaptes auratus ssp.	Northern (gilded) flicker	6.3		x	[30]
	Colaptes auratus ssp.	Common (red-shafted) flicker	7.6		x	[78]
	Dryocopus pileatus	Pileated woodpecker	9.0		x	[19] [67]
	Dryocopus pileatus	Pileated woodpecker	9.9		f	[30]
	Dryocopus pileatus	Pileated woodpecker	9.9	9.4	x	[78]
	Melanerpes aurifrons	Golden-fronted woodpecker	5.7		m	[30]
	Melanerpes carolinus	Red-bellied woodpecker	12.1		m	[30]
	Melanerpes erythrocephalus	Red-headed woodpecker	9.9		x	[30]
	Melanerpes formicivorus	Acorn woodpecker	16.0		m	[80]
	Melanerpes formicivorus	Acorn woodpecker	15.0		f	[80]
	Melanerpes uropygialis	Gila woodpecker	7.8		m	[30]
	Picoides borealis	Red-cockaded woodpecker	13.8		f	[30]
	Picoides borealis	Red-cockaded woodpecker (+)	12.0		x	[69]
	Picoides borealis	Red-cockaded woodpecker		13.0	f	[69]
	Picoides pubescens	Downy woodpecker	11.4		m	[30]
	Picoides scalaris	Ladder-backed woodpecker	4.5		m	[30]
	Picoides villosus	Hairy woodpecker	15.8		f	[30]
	Sphyrapicus varius	Yellow-bellied sapsucker	6.8		m	[30]
Podicipediformes						
Podicipedidae	Aechmophorus accidentalis	Western grebe	6.6		m	[78]
	Aechmophorus occidentalis	Western grebe	14.0		x	[41]
	Podiceps auritus	Horned grebe	5.2		x	[29]
	Podiceps grisegena	Red-necked grebe	4.8		x	[78]
	Podiceps nigricollis	Eared grebe	5.2		x	[29]
	Tachybaptus ruficollis	Little grebe	13.0		x	[37]
Procellariformes						
Diomedeidae	Diomedea albatrus	Short-tailed albatross	17.3		x	[29]
	Diomedea amsterdamensis	Amsterdam albatross	40.0		x	[37]
	Diomedea bulleri	Buller's albatross (+)	30.0		x	[37]
	Diomedea chlororhynchos	Yellow-nosed albatross	37.0		x	[37]
	Diomedea chrysostoma	Gray-headed albatross (+)	30.0		x	[37]
	Diomedea epomophora	Royal albatross	58.0		x	[37]
	Diomedea exulans	Wandering albatross	40.0		x	[37]
	Diomedea immutabilis	Laysan albatross	53.0		x	[45] [37]
	Diomedea immutablis	Laysan albatross	37.4		x	[29]
	Diomedea irrorata	Waved albatross	40.0		x	[37]
	Diomedea melanophris	Black-browed albatross	34.0		x	[37]
	Diomedea nigripes	Black-footed albatross	27.7		x	[29]
	Diomedea nigripes	Black-footed albatross	27.0		x	[58]
Hydrobatidae	Hydrobates pelagicus	European storm petrel (+)	20.0		x	[37]
	Oceanodroma furcata	Fork-tailed storm petrel	6.0		x	[29]
	Oceanodroma furcata	Fork-tailed storm petrel (+)	9.0		f	[78]

	Oceanodroma homochroa	Ashy storm petrel	8.0		x	[29]
	Oceanodroma leucorhoa	Leach's storm petrel	29.0		m	[29]
	Oceanodroma leucorhoa	Leach's storm petrel	31.0		x	[37]
	Oceanodroma leucorhoa	Leach's storm petrel (+)	24.0		m	[78]
	Oceanodroma tristrami	Sooty storm petrel	9.8		x	[29]
Procellariidae	Bulweria bulwerii	Bulwer's petrel	22.0		x	[29]
	Daption capense	Cape petrel	20.0		x	[37]
	Fulmarus glacialis	Northern fulmar	7.3		x	[29]
	Fulmarus glacialis	Northern fulmar	48.0		x	[37]
	Pagodroma nivea	Snow petrel	20.0		x	[37]
	Procellaria parkinsoni	Black petrel	17.0		x	[37]
	Pterodroma alba	Phoenix Island petrel	11.3		x	[29]
	Pterodroma hypoleuca	Bonin Island petrel	10.7		x	[29]
	Pterodroma leucoptera	Gould's petrel	20.0		x	[37]
	Pterodroma macroptera	Great-winged petrel (+)	18.0		x	[37]
	Pterodroma phaeopygia	Darked-rump petrel	7.0		x	[29]
	Puffinus auricularis	Townsend's shearwater (+)	5.1		x	[78]
	Puffinus lherminieri	Audobon's shearwater	19.0		x	[37]
	Puffinus nativitatis	Christmas Island shearwater	10.8		x	[29]
	Puffinus pacificus	Wedge-tailed shearwater	14.0		x	[29]
	Puffinus pacificus	Wedge-tailed shearwater	16.8		x	[78]
	Puffinus puffinus	Manx shearwater	4.8		x	[37]
	Puffinus puffinus	Manx shearwater	30.0		x	[78]
	Puffinus tenuirostris	Short-tailed shearwater	30.0		x	[37]
	Puffinus gravis	Greater shearwater	7.1		x	[29]
Psittaciformes						
Cacatuidae	Cacatua moluccensis	Rose-crested cockatoo		25.1	x	[50]
	Cacatua roseicapilla	Roseate cockatoo		26.3	x	[50]
	Calyptorhynchus banksii	Banksian cockatoo		29.3	x	[50]
Psittacidae	Amazona auropaleiata	Golden-naped parrot		49.0	x	[50]
	Ara ararauna	Blue and yellow macaw		43.0	x	[50]
	Ara chloroptera	Red and yellow macaw (+)		30.0	x	[50]
	Ara macao	Red and blue macaw		64.0	x	[50]
	Ara militaris	Military macaw		30.0	x	[50]
	Aratinga holochlora	Mexican conure		22.5	x	[50]
	Myiopsitta monacha	Quaker parakeet		17.8	x	[50]
	Nandayus nenday	Paraguayan black-headed conure		18.2	x	[50]
	Nestor notabilis	Kea		14.4	x	[50]
	Poicephalus meyeri transvaalensis	Brown parrot		34.2	x	[50]
	Poicephalus rufiventris	Red-breasted parrot		33.4	x	[50]
	Psittacus erithacus	African grey parrot		73.0	x	[50]
Sphenisciformes						
Spheniscidae	Aptenodytes patagonica	King penguin		26.0	x	[50]
	Aptenodytes patagonica	King penguin		23.7	x	[50]
	Spheniscus magellanicus	Magellanic penguin	13.0		x	[125]
Strigiformes						
Strigidae	Aegolius acadicus	Northern saw-whet owl	7.3		x	[26]
	Aegolius acadicus	Northern saw-whet owl	7.4		x	[30]

	Aegolius acadicus	Northern saw-whet owl	7.0	16.0	x	[78]
	Aegolius funereus	Boreal owl	3.4		x	[81]
	Aegolius funereus	Boreal owl	11.0		m	[60]
	Asio flammeus	Short-eared owl	12.9		x	[33]
	Asio otus	Long-eared owl	27.8		x	[30] [33]
	Athene cunicularia	Burrowing owl	8.7		x	[76]
	Bubo virginianus	Great horned owl	17.3		x	[30]
	Bubo virginianus	Great horned owl	20.6		x	[78]
	Micrathene whitneyi	Elf owl	4.9		x	[78]
	Nyctea scandiaca	Snowy owl	10.8		f	[78]
	Nyctea scandiaca	Snowy owl	9.4	28.0	x	[9] [123]
	Otus asio	Eastern screech-owl	14.2		x	[52]
	Otus flammeolus	Flammulated owl (+)	8.1		x	[119]
	Otus kennicottii	Western screech-owl	12.9		x	[30]
	Scotiaptex lapponica	Lap owl		27.0	x	[50]
	Strix nebulosa	Great gray owl (+)	13.0		f	[18]
	Strix nebulosa	Great gray owl	9.0		x	[105]
	Strix occidentalis	Spotted owl (+)	17.0	25.0	x	[56]
	Strix varia	Barred owl	10.1		f	[30]
	Strix varia	Barred owl	18.2		x	[78]
Tytonidae	Tyto alba	Common barn-owl	15.4		x	[30]
	Tyto alba	Barn owl		13.0	x	[50]
Struthioniformes						
Struthionidae	Struthio camelus	Ostrich		50.0	x	[82]
	Struthio camelus molybdophanes	Somali ostrich		26.1	x	[50]
Tinamiformes						
Tinamidae	Tinamus solitarius	Solitary tinamou		15.0	x	[37]
Trochiliformes						
Trochilidae	Archilochus colubris	Ruby-throated hummingbird	6.3		f	[30]
	Archilochus colubris	Ruby-throated hummingbird	9.0		x	[78]
	Calothorax lucifer	Lucifer hummingbird	4.0		x	[126]
	Calypte anna	Anna's hummingbird	6.0		f	[30]
	Calypte anna	Anna's hummingbird	6.3		x	[78]
	Lamprolaima rhami	Garnet-throated hummingbird(+)	8.0		m	[31]
	Selasphorus platycercus	Broad-tailed hummingbird	7.1		m	[23]
	Selasphorus platycercus	Broad-tailed hummingbird	12.2		f	[22]
	Selasphorus platycercus	Broad-tailed hummingbird	12.1		f	[30]
	Selasphorus platycercus	Broad-tailed hummingbird	8.0		x	[78]
	Selasphorus rufus	Rufous hummingbird	4.1		x	[78]
	Selasphorus sasin	Allen's hummingbird	3.9		f	[30]
	Selasphorus sasin	Allen's hummingbird	4.0		x	[78]
	Stellula calliope	Calliope hummingbird	6.1		f	[24]
	Stellula calliope	Calliope hummingbird	6.0		x	[78]

References

1. Amman, G.A., 1938. *The Life History and Distribution of the Yellow-headed Blackbird.* Ann Arbor, MI: University of Michigan.
2. Austin-Jr., O.L., 1968. Life histories of North American cardinals, grosbeaks, buntings, towhees, finches, sparrows and allies: part 1. *U. S. National Museum Bulletin.*
3. Basket, T.S. and M.W. Sayre, 1993. Characteristics and importance, in *Ecology and Management of the Morning Dove*, T.S. Basket, M.W. Sayre, R.E. Tomlinson, and R.E. Mirarchi, Editors. Harrisburg, PA: Stackpole Books. 1-6.
4. Bednarz, J.C., 1994. *Parabuteo unicinctus,* Harris' hawk, in *The Birds of North America*, A. Poole and F. Gill, Editors. Philadelphia: The American Ornithologist's Union & The Academy of Natural Sciences of Philadelphia. 4(146).
5. Bennett, E.T., 1829. *The Tower Menagerie*. London.
6. Bennetts, R.E. and W.M. Kichens, 1993. Estimation and environmental correlates of survival and dispersal of snail kites in Florida. *U. S. Fish and Wildlife Service and U.S. National Park Service Florida Corporation of Fish Wildlife Reserve Unit.*
7. Bergstrom, E.A., 1951. Some recoveries of black crowned night herons. *Bird Banding* 22: 80-81.
8. Binkley, C.S. and R.S. Miller, 1983. Population characteristics of the whooping crane: *Grus americanaCanadian Journal of Zoology* 61: 2768-2776.
9. Blotzheim, U.N.Glutz von and K.M. Bauer, eds. 1980. *Handbuch der Vogel Mitteleuropas*. Blotzheim, U.N.G.v. and K.M. Bauer, Eds. Vol. 9. Wiesbaden: Akad. Verlag.
10. Boag, D. and M. Schroeder, 1992. *Dendragapus canadensis*, spruce grouse, in *The Birds of North America*, A. Poole and F. Gill, Editors. Philadelphia: The American Ornithologist's Union & The Academy of Natural Science of Philadelphia. 1 (5):17.
11. Bollinger, E.K., 1988. A longevity record for the Bobolink. *North American Bird Bander* 13: 76.
12. Bowers, R.K.J. and J.B.J. Dunning, 1994. *Epidomax fulvifrons,* buff-breasted flycatcher, in *The Birds of North America*, A. Poole and F. Gill, Editors. Philadelphia: The American Ornithologists Union & The Academy of Natural Sciences of Philadelphia. 4(125).
13. Braun, C.E., K. Martin, and L.A. Robb, 1993. *Lagopus leucurus,* white-tailed ptarmigan, in *The Birds of North America*, A. Poole and F. Gill, Editors. Philadelphia: The American Ornithologist's Union & The Academy of Natural Science of Philadelphia. 2 (68): 12.
14. Briskie, J.V., 1992. *Calcarius pictus*, Smith's longspur, in *The Birds of North America*, A. Poole and F. Gill, Editors. Philadelphia: The American Ornithologist's Union & The Academy of Natural Science of Philadelphia. 1(34): 13.
15. Brown, C.R. and M.B. Brown, 1994. *Hirundo pyrrhonota,* cliff swallow, in *The Birds of North America*, A. Poole and F. Gill, Editors. Philadelphia: The American Ornithologist's Union & The Academy of Natural Sciences of Philadelphia. 4(149)

16. Brown, J.L., 1994. *Aphelocoma ultramarina,* Mexican jayin *The Birds of North America*, A. Poole and F. Gill, Editors. Philadelphia: The American Ornithologist Union & The Academy of Natural Sciences of Philadelphia. 3(118): 12.
17. Brownie, C., D.R. Anderson, K.P. Burnham, and D.S. Robson, eds. 1985. *Statistical Inference From Band Recovery Data-A Handbook.* 2 ed. U. S. Fish Wildlife Service, Resource Publication.
18. Bull, E.L. and J.R. Duncan, 1993. *Strix nebulosa,* great grey owl, in *The Birds of North America*, A. Poole and F. Gill, Editors. Philadelphia: The American Ornithologist's Union & The Academy of Natural Science of Philadelphia. 2(41): 11.
19. Bull, E.L. and J.A. Jackson, 1994. *Dryocopus pileatus,* pileated woodpecker, in *The Birds of North America*, A. Poole and F. Gill, Editors. Philadelphia: The American Ornithologist's Union & The Academy of Natural Sciences of Philadelphia. 4(148): 14.
20. Butler, R.W., 1988. Population dynamics and migration routes of tree swallows, *Tachycineta bicolor*, in North America. *Journal of Field Ornithology* 59: 395-402.
21. Byrd, G.V. and J.C. Williams, 1993. *Rissa brevirostris,* red-legged kittiwake, in *The Birds of North America*, A. Poole and F. Gill, Editors. Philadelphia: The American Ornithologist's Union & The Academy of Natural Science of Philadelphia. 2(60): 7.
22. Calder, W.A., 1990. Avian longevity and aging, in *Genetic Effects on Aging II*, D.E. Harrison, Editor. Caldwell NJ 185-204.
23. Calder, W.A. and L.L. Calder, 1992. *Selasphorus platycercus*, broad-tailed hummingbird, in *The Birds of North America*, A. Poole and F. Gill, Editors. Philadelphia: The American Ornithologist's Union & The Academy of Natural Science of Philadelphia. 1(16): 10.
24. Calder, W.A. and L.L. Calder, 1994. *Stellula calliope*, calliope hummingbird, in *The Birds of North America*, A. Poole and F. Gill, Editors. Philadelphia: The American Ornithologist's Union & The Academy of Natural Sciences of Philadelphia. 4(135).
25. Campbell, R.W., N.K. Daw, I. McTaggart-Cowan, J.M. Cooper, G.W. Kaiser, and M.C.E. McNall, eds. 1990. *The Birds of British Columbia*. Campbell, R.W., N.K. Daw, I. McTaggart-Cowan, J.M. Cooper, G.W. Kaiser, and M.C.E. McNall, Eds. Vol. 2. Vancouver, BC: Mitchell Press.
26. Cannings, R.J., 1993. *Aegolius acadicus*, northern saw-whet owl, in *The Birds of North America*, A. Poole and F. Gill, Editors. Philadelphia: The American Orithologist's Union & The Academy of Natural Science Of Philadelphia. 2(42): 11.
27. Carrol, J.P., 1993. Gray partridge (*Perdix perdix*), in *The Birds of North America*, A. Poole and F. Gill, Editors. Philadelphia: The American Ornithologist's Union & The Academy of Natural Sciences. 2(58)
28. Carter, H.R., G.J. McChesney, D.C. Jaques, C.S. Strong, M.W. Parker, J.E. Takekawa, D.L. Tory, and D.L. Whitworth, 1992. *Breeding Populations of Seabirds in California, 1989-1991, Vol. 1: Population Estimates*. Dixon, CA: U.S. Fish and Wildlife Service, Northern Prairie Wildlife Research Center.
29. Clapp, R.B., M.K. Klimkiewicz, and J.H. Kennard, 1982. Longevity records of North American birds: Gaviidae through Alcidae. *Journal of Field Ornithology* 53: 81-124, 125-208.

30. Clapp, R.B., N.K. Klimkiewicz, and A.G. Futcher, 1983. Longevity records of North American birds: Columbidae through Paridae. *Journal of Field Ornithology* 54: 123-137.
31. Conway, W.G., 1961. Hummingbirds with wrinkles. *Animal Kingdom* 64: 151.
32. Cooke, M.T., 1942. Returns from banded birds--some longevity records of banded birds. *Bird Banding* 13: 110-119.
33. Cramp, S., 1985. *The Birds of the Western Palearctic* Vol. 4. Oxford: Oxford University Press.
34. Crandall, L.S., 1952. Unpublished data from the New York Zoological Society.
35. Csada, R.D. and M. Brigham, 1992. *Phalaenoptilus nuttallii*, common poorwill, in *The Birds of North America*, A. Poole and F. Gill, Editors. Philadelphia: The American Ornithologist's Union & The Academy of Natural Sciencs of Philadelphia. 1, Number 32 p. 10.
36. Davis, J.N., 1995. *Vireo huttoni*, Hutton's vireo, in *The Birds of North America*, A. Poole and F. Gill, Editors. Philadelphia, PA: The American Orthinologists' Union and The Academy of Natural Sciences of Philadelphia. 5(189): 13.
37. delHoyo, J., A. Elliot, and J. Sargatal, eds. 1992. *Handbook of Birds of the World, Vol. 1*. Barcelona: Lynx Edicions.
38. Dugger, B.D., K.M. Dugger, and L. H.Fredrickson, 1994. *Lophodytes cucullatus,* hooded merganser, in *The Birds of North America*, A. Poole and F. Gill, Editors. Philadelphia: The American Ornithologist's Union & The Academy of Natural Science. 3(98): 12.
39. Dunning-Jr., J.B., 1992. Significant encounters. *North American Bird-Bander* 17: 29.
40. Eadie, J.M., M.L. Mallory, and A.G. Lumsden, 1995. *Bucephala clangula*, common goldeneye, in *The Birds of North America*, A. Poole and F. Gill, Editors. Philadelphia, PA: The American Orthinologists' Union and The Academy of Natural Sciences of Philadelphia. 5(170): 12.
41. Eichhort, B.A., 1992. An analysis of western grebe banding and recovering data. *North American Bird Bander* 17: 108-115.
42. Ewins, P.J., H.R. Carter, and V.S. Yu, 1993. The status, distribution and ecology of inshore fish-feeding alcids (*Cepphus* guillemots and *Brachyramphus* murrelets) in the North Pacific, in *Status, Ecology, and Conservation of Marine Birds of the North Pacific*, K.Vermeer., ed. Special Publication Canadian Wildlife Service and Pacific Seabird Group.
43. Feare, C.J., ed. 1984. *The Starling*. Oxford: Oxford University Press.
44. Ficken, M. and J. Nocedal, 1992. *Parus sclateri*, Mexican chickadee, in *The Birds of North America*, A. Poole and F. Gill, Editors. Philadelphia: The American Ornithologist's Union & The Academy of Natural Science of Philadelphia. 1(8): 8.
45. Fisher, H.I., 1975. Longevity of the Laysan albatross, *Diomedia immutabius. Bird Banding* 46: 1-6.
46. Fitzinger, L.J.F.J., 1853. Versuch einer Geschicte der Menagerien des Osterreichisch-Kaiserlichen Hofes. *Aus dem Marz-und Aprilhefte des Jahrganges 1853 der Sitzungsberichte der mathem.- naturw. Classe der kaie. Akademie der Wissenschaften besonders abgedruckt.*

47. Flores, R.E. and W.R. Eddleman, 1991. *Ecology of the California Black Rail in Southwestern Arizona: Final Report*. Yuma , AZ: U.S. Bureau Reclamation, Yuma Project Office & Arizona Department Game & Fish.
48. Flower, M.S.S., 1925. Contributions to our knowledge of the duration of life in vertebrate animals - III. Reptiles. *Proceedings of the Zoological Society of London* 1925(2): 911-981.
49. Flower, M.S.S., 1925. Contributions to our knowledge of the duration of life in vertebrate animals - IV. Birds. *Proceedings of the Zoological Society of London* 1925(2): 1365-1422.
50. Flower, M.S.S., 1938. The duration of life in animals - IV. Birds: special notes by orders and families, in *Proceedings of the Zoological Society of London*. 195-235.
51. Gaston, A.J., 1994. *Synthliboramphus antiques*, ancient murrelet, in *The Birds of North America*, A. Poole and F. Gill, Editors. Philadelphia: The American Ornithologist's Union & The Academy of Natural Science of Philadelphia. 4 (132).
52. Gehlbach, F.R., 1994. *The Eastern Screech-Owl: Life History, Ecology, and Behavior in Suburbia and the Countryside*. College Station, TX: Texas A & M University Press.
53. Goodpasture, K.A., 1977. Returns records of Louisiana waterthrushes: an eight-year-old bird reported. *Bird Banding* 48: 152-154.
54. Greenlaw, J.S. and J.D. Rising, 1994. *Ammodramus caudatus,* sharp-tailed sparrow, in *The Birds of North America*, A. Poole and F. Gill, Editors. Philadelphia: The American Ornithologists Union & The Academy of Natural Sciences of Philadelphia. 3 (112): 17.
55. Grzybowski, J.A., 1995. *Vireo atricapillus*, black-capped vireo, in *The Birds of North America*, A. Poole and F. Gill, Editors. Philadelphia, PA: The American Ornithologists' Union and The Academy of Natural Sciences of Philadelphia. 5(181):18.
56. Gutierrez, R.J., A.B. Franklin, and W.W. Lahaye, 1995. *Strix occidentalis*, (Northern) spotted owl, in *The Birds of North America*, A. Poole and F. Gill, Editors. Philadelphia, PA: The American Ornithologists' Union and The Academy of Natural Sciences of Philadelphia. 5(179): 15.
57. Haggerty, T.M., 1988. Aspects of the breeding biology and productivity of Bachmans sparrow in central Arkansas. *Wilson Bulletin* 100: 247-255.
58. Harrison, C.S., ed. 1990. *Seabirds of Hawaii*. Ithaca, NY: Comstock/ Cornell.
59. Hatch, S.A., G.V. Byrd, D.B. Irons, and G.L. Hunt, 1993. Status and ecology of Kittiwakes (*Rissa tridactyla* and *Rissa brevirostris*) in the North Pacific, in *The Status, Ecology and Conservation of Marine Birds in the North Pacific*, K. Vermeer, K.T. Briggs, K.H. Morgan, and D. Siegel-Causey, Editors. Ottawa, Ontario: Canadian Wildlife Service Special Publications 140-153.
60. Hayward, G.D. and P.H. Hayward, 1993. *Aegolius funereus*, boreal owl, in *The Birds of North America*, A. Poole and F. Gill, Editors. Philadelphia: The American Ornithologist's Union & The Academy of Natural Science of Philadelphia. 2(63):14.
61. Holden, G.H., 1883. *Canaries and Cage-birds*. New York: G. H. Holden.
62. Holland, P., 1992. Recent recoveries of waders. *Wader Study Group Bulletin* 64: 63-64.
63. Holroyd, G.L. and R. Brown, 1970. Longevity record of semipalmated sandpiper. *Ontario Bird Banding* 6: 73.

64. Houston, C.S., 1974. Longevity record for black-crowned night heron: 16 1/2 Years. *Blue Jay* 32: 179.

65. Houston, C.S., 1984. Unusual story record 20-year longevity of ferriginous hawk. *Blue Jay* 42: 99-101.

66. Houston, M.I. and C.S. Houston, 1987. Tree swallowbanding near Saskatoon, Saskatchewan. *North American Bird Bander* 12: 103-108.

67. Hoyt, J.S.Y. and S.F. Hoyt, 1951. Age records of pileated woodpeckers. *Bird Banding* 22: 125.

68. Hussel, D.J.T., 1982. Longevity and fecundity records in the tree swallow. *North American Bird Bander* 12: 103-108.

69. Jackson, J.A., 1994. *Picoides borealis*, red-cockaded woodpecker, in *The Birds of North America*, A. Poole and F. Gill, Editors. Philadelphia: The American Ornithologist's Union & The Academy of Natural Science of Philadelphia. 3(85):11.

70. Johnson, S.R. and R.W. Campbell, 1994. *Acridotheres cristatellus*, crested myna, in *The Birds of North America*, A. Poole and F. Gill, Editors. Philadelphia: The American Ornithologist's Union & The Academy of Natural Science of Philadelphia. 4(157): 11.

71. Johnston, R.F., 1992. *Columba livia*, rock dove, in *The Birds of North America*, A. Poole and F. Gill, Editors. Philadelphia: The Ornithologist's Union & The Academy of Natural Science of Philadelphia. 1(13): 9.

72. Jones, I.L., 1992. Factors affecting adult survival of least auklets at St. Paul Island, Alaska. *Auk* 109: 576-584.

73. Jones, I.L., 1993. *Aethia cristatella*, crested Auklet, in *The Birds of North America*, A. Poole and F. Gill, Editors. Philadelphia: The American Ornithologist's Union & The Academy of Natural Science of Philadelphia. 2(70): 11.

74. Jones, I.L., 1993. *Aethia pusilla,* least auklet, in *The Birds of North America*, A. Poole and F. Gill, Editors. Philadelphia: The American Ornithologist's Union & the Academy of Natural Science of Philadelphia. 2(69): 11.

75. Kemp, A., 1995. *The Hornbills*. New York: Oxford University Press.

76. Kennard, J.H., 1975. Longevity records of North American birds. *Bird Banding* 46: 55-59, 59-73.

77. Klimkiewicz, M.K., R.B. Clapp, and A.G. Futcher, 1983. Longevity records of North American birds: Remizidae through Parulinae. *Journal of Field Ornithology* 54: 287-294.

78. Klimkiewicz, M.K. and A.G. Futcher, 1989. Longevity records of North American birds: supplement 1. *Journal of Field Ornithology* 60: 469-494.

79. Klimkiewicz, M.K. and A.G. Futcher, 1987. Longevity records of North American birds: Coerbinae through Estrildidae. *Journal of Field Ornithology* 58: 318-333.

80. Koenig, W.D., P.B. Stacey, M.T. Stanback, and R.L. Mumms, 1995. *Melanerpes formicivorus,* acorn woodpecker, in *The Birds of North America*, A. Poole and F. Gill, Editors. Philadelphia, PA: The American Orthinologists' Union and The Academy of Natural Sciences of Philadelphia. 5(194):17.

81. Korpimaki, E., 1988. Effects of age on breeding performance of Tengmalin's owl *Aegolius funerous* in western Finland. *Ornis. Scand.* 19.

82. Korschelt, E., 1922. *Lebensdauer Altern und Tod*. Jena: G. Fischer.

83. Kortright, F.H., 1943. *The Ducks, Geese, & Swans of North America*. Washington D.C.: American Wildlife Institute.

84. Kuhk, R., 1956. Hagelunwetter als Verlustursache Bei Strochen und Andren Vogeln. *Vogelwarte* 19: 145.

85. Kurt, E. and J.P. Goosen, 1987. Survival, age composition, sex ratio, and age at first breeding of whooping cranes in Wood Buffalo National Park, Canada, in *Proceedings 1985 International Crane Workshop*, J.C. Lewis and J.W. Ziewitz, Editors. Grand Island, NE: Platte River Whooping Crane Habitat Main. Trust and U. S. Wildlife Service. 230-244.

86. Lanyon, W.E., 1979. Hybrid sterility in meadowlarks. *Nature* 279: 557-558.

87. Leberman, R.C., R.S. Mulvihill, and D.S. Wood, 1985. Bird-banding at Powdermill, 1983, in *Powdermill Nature Reserve Research Report*. Pittsburgh: Carnegie Museum Natural History. 44 p.

88. Limpert, R.J. and S.L. Earnst, 1994. *Cygnus columbianus*, tundra swan, in *The Birds of North America*, A. Poole and F. Gill, Editors. Philadelphia: The American Ornithologist's Union & The Academy of Natural Science of Philadelphia. 3(89): 13.

89. Lloyd, H., 1937. Twenty-year old ferriginous rough-legged hawk. *Canadian Field Naturalist* 51: 137.

90. Mann, W.M., 1930. Wild animals in and out of the zoo, in *Smithsonian Scientific Series* 6:30-337.

91. Manuwal, D.A. and A.C. Thoresen, 1994. *Ptychoramphus aleuticus*, cassin's auklet, in *The Birds of North America*, A. Poole and F. Gill, Editors. Philadelphia: The American Ornithologist's Union & The Academy of Natural Science of Philadelphia. 2(50):12.

92. Martin, T.E. and P.M. Barber, 1994. *Cardellina rubifrons*, red-faced warbler, in *The Birds of North America*, A. Poole and F. Gill, Editors. Philadelphia: The American Ornithologist's Union & The Academy of Natural Science of Philadelphia. 4(152): 10.

93. Mayfield, H.F., ed. 1960. *The Kirtland's Warbler*. Mayfield, H.F., Bloomfield Hills, Michegan: Cranbrook Institute Science.

94. Middleton, A.L.A., 1984. Longevity of the American goldfinch. *Journal of Field Ornithology* 55: 383-386.

95. Miller, E.H. and R. McNeil, 1988. The longevity record for the least sandpiper: a revision. *Journal of Field Ornithology* 59: 403-404.

96. Mills, S., J. Silliman, K. Groschupf, and S. Speich, 1980. Life history of the five-striped sparrow. *Living Bird* 18: 95-110.

97. Mirande, C.R.L., ed. 1993. *Whooping Crane (Grus americana) Consultation Viability Assessment Workshop Report*. Mirande, C.R.L., Apple Valley, MN: Captive Breeding Specialist Group, International Union For Conservation of Nature.

98. Mitchell, P.C., 1911. *Proceedings of the Zoology Society of London* 1911: 425.

99. Moltoni, E., 1947. Uccisione di una Tortora del Collare orientale, *Streptopelia d. decato (*Fridvalsky*)* in quel di caorle (venezia). *Rivue Italia Ornithologia* 17: 139.

100. Moody, A.F., 1931. Death of an American whooping crane *Aviculture Magazine,* Vol. 9: 8-11:

101. Newton, I. and P. Olsen, eds. 1990. *Birds of Prey*. New York: Facts on File, Inc.

102. Nice, M.M., 1966. Territory defense in the oystercatcher. *Bird Banding* 37: 132.

103. Nickell, W.P., 1966. The nesting of the black-crowned night heron and its associates. *Jack Pine Warbler* 32: 179.

104. Nol, E. and R.C. Humphrey, 1994. *Haematopus palliatus*, American oystercatcher, in *The Birds of North America*, A. Poole and F. Gill, Editors. Philadelphia: The American Ornithologist's Union & The Academy of Natural Science. 3(82):13.

105. Oeming, A.F., 1964. Banding recovery of great gray owl. *Blue Jay* 22: 10.

106. Oliphant, L.W., I.G. Warkentin, N.S. Sodhi, and P.C. James, In Press. Ecology of urban merlins in Saskatoon, in *Biology and Conservation of Small Falcons*, R. Chancellor, R. Clark, S. Dewer, and M. Nicholls, Editors. Canterbury, United Kingdom: Hawk and Owl Trust.

107. Page, G.W., J.S. Warriner, J.C. Warriner, and P.W.C. Paton, 1994. *Charadrius alexandrinus*snowy plover, in *The Birds of North America*, A. Poole and F. Gill, Editors. Philadelphia: The American Ornithologist's Union & The Academy of Natural Science of Philadelphia. 4(154): 15.

108. Palmer, R.S., ed. 1988. *Red-tailed Hawk*. Handbook of North American Birds. Palmer, R.S., Eds. Vol. 5. New Haven, CT: Yale University Press.

109. Paludan, K., 1963. Partridge markings in Denmark. *Danish Review of Game Biology* 4: 25-60.

110. Payne, R.B. and L. L.Payne, 1990. Survival estimates of indigo buntings: Comparison of banding recoveries and local observations. *Condor* 92: 938-946.

111. Peterson, R.W., 1971. Warbler returns at Somesville, Maine. *Bird Banding* 42: 99-102.

112. Picchi, C., 1913. Observations sur la longevite des oiseaux. *Bulletin of the Society of Zoologists* 38: 212.

113. Pierotti, R.J. and C.A. Annett, 1994. *Larus occidentalis*, western gull, in *The Birds of North America*, A. Poole and F. Gill, Editors. Philadelphia, PA: The American Orthinologists' Union and The Academy of Natural Sciences of Philadelphia. 5(174): 16.

114. Post, W. and J.S. Greenlaw, 1994. *Ammodramus maritmus* seaside sparrow, in *The Birds of North America*, A. Poole and F. Gill, Editors. Philadelphia: The American Ornithologist's Union & The Academy of Natural Sciences of Phiadelphia. 4(127): 18.

115. Purdy, M.A., 1985. *Parental Behavior and Role Differentation in the Black Oystercatcher Haematopus bachmani*. Victoria: University of Victoria.

116. Rankin, N., 1957. Longevity of a white-fronted goose. *British Birds* 50: 164.

117. Reynolds, M.D., 1995. *Pica nuttalli*, Yellow-billed magpie, in *The Birds of North America*, A. Poole and F. Gill, Editors. Philadelphia, PA: The American Orthinologists' Union and The Academy of Natural Sciences of Philadelphia. 5(180):15.

118. Reynolds, R.T. and B.D. Unkhart, 1990. Longevity records for male and female flammulated owls. *Journal of Field Ornithology* 61: 243-244.

119. Robel, R.J. and J. W. B. Ballard, 1974. Lek social organization and reproductive success in the greater prairie chicken. *American Zoology* 14: 121-128.

120. Rosenfield, R.N. and J. Bielefeldt, 1993. *Accipiter cooperii*, Cooper's Hawk, in *The Birds of North America*, A. Poole and F. Gill, Editors. Philadelphia: The American Ornithologist's Union. 2, Number 75 p. 14.
121. Ryder, J.P. and R.T. Alisauskas, 1995. *Chen rossii*, Ross's goose, in *The Birds of North America*, A. Poole and F. Gill, Editors. Philadelphia, PA: The American Orthinologist's Union and The Academy of Natural Sciences of Philadelphia. 5(162): 16.
122. Rydzewski, W., 1978. The Longevity of Ringed Birds. *Ring* 96-97: 218-262.
123. Schenker, A., 1978. Hochsalter Europaischer vogel im Zoologischen Garten Basel. *Ornithol. Beob.* 75: 96-97.
124. Schnell, J.H., 1994. *Buteogallus anthracinus*, common black hawk, in *The Birds of North America*, A. Poole and F. Gill, Editors. Philadelphia: The American Ornithologist's Union & The Academy of Natural Science of Philadelphia. 4(122).
125. Scolario, J.A., 1990. On a longevity record of the Magellanic penguin. *Journal of Field Ornithology* 61: 377-484.
126. Scott, P.E., 1994. Lucifer hummingbird (*Calothorax lucifer*), in *The Birds of North America*, A. Poole and F. Gill, Editors. Philadelphia: The Academy of Natural Sciences & The American Ornithologist's Union. 4(138).
127. Sedgwick, J.A., 1993. *Empidonax oberholseri*, dusky flycatcher in *The Birds of North America*, A. Poole and F. Gill, Editors. Philadelphia: The American Ornithologist's Union & The Academy of Natural Science of Philadelphia. 2(78): 12.
128. Smith, S.M., 1991. *The Black-capped Chickadee: Behavioral Ecology and Natural History*. Ithaca, NY: Cornell University Press.
129. Southern, W.E., 1975. Longevity records for ring-billed gulls. *Auk* 92: 369.
130. Spil, R.E., M.W.V. Walstijn, and H. Albrecht, 1985. Observations on the behavior of the scarlet ibis, *Eudocimus ruber*, in Artis Zoo, Amsterdam. *Bijdr. Dierkd.* 55: 219-232.
131. Strickland, D. and H. Quellet, 1992. *Perisreus canadensis*, gray jay, in *The Birds of North America*, A. Poole and F. Gill, Editors. Philadelphia: The American Ornithologist's Union & The Academy of Natural Science of Philadelphia. 1(40): 17.
132. Stutzenbaker, C.D., ed. 1988. *The Mottled Duck, its Life History, Ecology and Management*. Austin: Texas Parks Wildlife Department.
133. Sykes-Jr., P.W., 1980. Decline and disappearance of the dusky seaside sparrow from Merrit Island, Florida. *American Birds* 34: 728-737.
134. Tacha, T.C., S.A. Nesbitt, and P.A. Vohs, 1992. *Grus canadensis* sandhill crane, in *The Birds of North America*, A. Poole and F. Gill, Editors. Philadelphia: The American Ornithologist's Union & The Academy of Natural Science of Philadelphia. 1(31): 15.
135. Tarboton, W., 1989. An old cattle egret. *Witwaterstrand Bird Club Newsletter* 147: 10.
136. Terres, J.K., 1968. *Flashing Wings: The Drama of Bird Flight*. Garden City, New York: Doubleday.
137. Thomas, V.G. and S.D. MacDonald, 1987. The breeding distribution and current population status of the ivory gull in Canada. *Arctic* 40: 211-218.
138. Vire, F., 1956. *Proceedings of the Verb. Society Science Natural Tunis* 4-5: 37.

139. Virkkala, R., 1990. Ecology of the Siberian tit *Parus cinctus* during the non-breeding season. *Ornis. Fenn.* 67: 98-99.

140. Walkingshaw, L.H., ed. 1983. *Kirtland's Warbler: The Natural History of an Endangered Species*. Walkingshaw, L.H., Bloomfield Hills, Michigan: Cranbrook Institute Science.

141. Weitnauer, E., 1960. *Ornithol. Beob.* 57: 158.

142. West, S., 1994. *Hirundo fulva,* cave swallow, in *The Birds of North America*, A. Poole and F. Gill, Editors. Philadelphia: The American Ornithologist's Union & The Academy of Natural Sciences of Philadelphia. 4(141): 12.

143. Weston, H.G., 1947. Breeding behavior of the blackheaded grosbeak. *Condor* 49: 54-73.

144. Wilcox, L., 1959. A twenty year banding study of the piping plover. *Auk* 76: 129-152.

145. Williams-Jr., L.E. and D.H. Austin, 1988. Studies of the wild turkey in Florida. *Technical Bulletin* 10.

146. Wilson, W.H., 1994. *Calidris mauri*, western sandpiper, in *The Birds of North America*, A. Poole and F. Gill, Editors. Philadelphia: The American Ornithologist's Union & The Academy of Natural Science of Philadelphia. 3(90):13.

147. Zwickel, F.C., D.A. Boag, and J.F. Bendall, 1989. Longevity in blue grouse. *North American Bird Bander* 14: 1-4.

Amphibians and Reptiles

Amphibians

Approximately 4,400 species are included in the three orders of amphibians. About 3% of the species are caecilians (Gymnophiona), 12% are salamanders (Caudata) and 85% are frogs (Anura). Amphibians are the only vertebrates to have a free-living aquatic developmental stage and a terrestrial juvenile/adult stage. Amphibians rely on cutaneous respiration and two types of skin glands, have a double auditory system, and specialized visual cells in the retina.

Amphibians differ greatly in morphology between and within orders. Caecilians resemble earthworms while the salamanders resemble lizards; frogs are characterized by large hindlimbs built for jumping.

GYMNOPHIONA (caecilians)

Caecilians are limbless, earthworm-like vertebrates that are almost pan-tropical (excepting Madagascar and Papua-Australia; Zug 1993). Their habitat includes moist soil, abandoned termite mounds, decaying material, and underneath rocks (Obst 1988). The skull is compact and tightly knit to withstand the pressures of tunneling. The eyes are regressed under skin and sometimes skull. They have no ear openings, and a pair of retractable sensory tentacles and an under-hung jaw to identify and capture prey in the tunnels. Caecilian bodies are grooved and segmented; their tails are short or non-existent, and their skin bare (some species have scales below the surface). Movement in tunnels is swift in either a forward or backward direction. Caecilians surface after dark to feed on arthropods, earthworms, cockroaches and other invertebrates and possibly to mate. Some cannibalism also occurs (Obst 1988).

Fertilization is internal for all caecilians. Males have a large, reversible copulatory organ. Development can be either internal or external (Zug 1993). In egg laying species, the clutch of about 30 large eggs are attached to one another; the female curls up around the eggs, and those which are separated from the mother do not seem to develop normally (Obst 1988).

CAUDATA (Salamanders and newts)

There are about 390 extant species of salamanders and newts in eight or more families. Caudates are distributed almost exclusively in Northern Hemisphere temperate forests. Lifestyles range from exclusively aquatic to entirely terrestrial (Zug 1993). All have long tails, a cylindrical body and distinct heads. Protective adaptations include nocturnal behavior, poisonous secretions (usually accompanied by bright warning coloration), and breakable tails.

Fertilization is generally internal (except in Cryptobranchoidea), though the male lacks a copulatory organ. The male releases a spermophore (sperm filled gelatinous capsule) which the female then picks up in her cloaca. The female stores the sperm for days to years. Typically the eggs are deposited in water. Larvae are predominately predatory, eating little or no plant matter. Terrestrial species tend to lay larger eggs with young that hatch in miniature adult form (Obst 1988).

ANURA (frogs and toads)

Anurans are pandemic except for some oceanic islands and the Polar Regions and are most diverse in the tropics. Frogs are found in rainforests, deserts, lowlands, mountains, in water and far from water (Obst 1988). All extant frogs share compact frame, strongly developed hind limbs, no tail and, in most species, no ribs. They have flat, broad heads and large mouths. A large, anteriorly attached tongue is used to capture prey. Most males have vocal sacs used in territory maintenance and courting.

Fertilization is typically external through amplexus. The male grasps the female from above and sperm are deposited as eggs are shed. Most commonly, the egg and sperm are released into open water where fertilization occurs. Anuran larvae (tadpoles) differ from those of Caudata in having a respiratory tube (spiracle). External gills are only present immediately after hatching. Larvae of most species are herbivorous (Obst 1988).

Reptiles

Over 5500 species in four orders comprise the reptiles. Snakes and lizards (Squamata) make up over 5000 of these species. There are approximately 200 species of turtles (Testudinata). Characteristics shared by all reptiles include pulmonary respiration, a single occipital condyle, two sacral vertebrae, epidermal scales, internal fertilization, and shelled, amniotic eggs. There are two ancient lines of reptiles, the diapsids (the lizards, snakes, tuataras, crocodilians, and sometimes birds) and the anapsids (turtles).

CROCODYLIA (crocodiles, alligators, caimans)

There are three extant families of Crocodylians; about 14 known fossil families extend back to the Triassic. They occur in the tropics and subtropics, in or near freshwater. Crocodylians have powerful tails same length or longer than their bodies, and are excellent swimmers using an undulating motion of the tail to propel themselves. Crocodylians rely on stealth and ambush to capture invertebrates, small fish, birds, and mammals. Ranging in size from 2 to 7 m or more, the larger Crocodylians have no enemies except humans (Obst 1988; Zug 1993).

Crocodylians are the only heterothermic reptiles with a fully developed secondary palate and a four chambered heart (Zug 1993). A large lung capacity and low metabolism allows them to stay submerged for more than an hour. Webbed hind limbs aid in swimming; however, Crocodylians are also highly adept at walking or running on land. Crocodylian skin is thick with large armor-like scales. They have strong jaws and angled teeth that are ideal for tightly gripping prey (Obst 1988).

Crocodylians lay clutches of 15 to 100 eggs and incubation lasts 64 to 115 days at constant temperatures of 29 to 34 degrees Celsius and humidity near 100%. Females guard the nest and aid the young in hatching from the eggs (Obst 1988). With low predation, crocodylians may attain natural ages of up to 100 years.

TESTUDINES (turtles and tortoises)

Modern turtles first appeared in the Upper Cretaceous and have remained basically unchanged since that time (Obst 1988). Distributed throughout all warm and most temperate areas, they can be found in varied habitats, including damp terrestrial, dry steppes and deserts, rivers and lakes, swamps, and marine habitats. The carapace (dorsal shell) contains dermal bones spread over and fused to the trunk vertebrate and ribs. The plastron (ventral shell) is an expressed plate of dermal plates with a few elements of the pectoral girdle and sternum. Bony ridges or ligaments connect the carapace and the plastron. In most extant turtles, the limbs, head and neck can be drawn inside the shell (Zug 1993, Obst 1988).

Turtles lack teeth but have a keratinous jaw-sheath that grows continuously. Various species of turtles may be herbivores, carnivores, or omnivores. Their necks have eight cervical vertebrae allowing high mobility and retractability. Limbs are highly developed for the environment; fin or flipper-like limb structures for aquatic species and stump-like limbs in terrestrial species. The rigidity of the carapace results in ribs that are not able to move; instead, muscles push air in and out of the lungs. Some aquatic species have also developed cutaneous respiration and gas exchange (Obst 1988).

All turtles are oviparous. There are no live-bearing forms, and all presumably lay eggs at a fairly early stage of development. Females can store sperm, delaying fertilization for up to four years. Nesting is uniform in turtles. Regardless of hindfoot morphology, they dig an egg pit with an alternating scooping motion of the hindfeet (Zug 1993).

RYNCHOCEPHALIA (tuataras)

Today there are only two extant species of tuatara (family Sphenodontida). Tuataras are native to roughly 30 small islands off the New Zealand coast where they have few predators but are able to prey on nesting seabirds. Although tuataras are most active at night, they are not entirely nocturnal, taking a wide variety of prey including insects, skinks, geckos, and hatchling seabirds (Zug 1993). They have a stout, lizard-like body, a large head, thick tail, and free abdominal ribs (Obst 1988). Mating of tuataras occurs in January, but egg laying is delayed until October-December. Females dig a nest cavity and deposit 8-15 eggs which will not hatch for 11 to 16 months (Zug 1993).

SQUAMATA (lizards, snakes, amphisbaenians)

Squamata are scaled reptiles such as lizards and snakes. They are distributed worldwide (Obst 1988). A recent study identified more than 70 shared traits of squamates in support of their monophyly including the paired hemipenes, which is unique among vertebrates. Also distinct for this order are the more or less regularly arranged horny scales that have to be shed periodically by molting (Obst 1988). Most squamates lay soft, parchment-like eggs that harden secondarily only in the Gekkonidae. A few squamates are ovoviviparous or viviparous (Obst 1988).

References Cited

Obst, F. J., 1988. *The Completely Illustrated Atlas of Reptiles and Amphibians for the Terrarium*. Neptune City, New Jersey, T. F. H. Publications.

Zug, G. R., 1993. *Herpetology: An Introduction of Amphibians and Reptiles*. San Diego, Academic Press.

Table 3. Record Life Spans (years) of Amphibians and Reptiles

Class Order/Family	Genus/Species/Sub-species	Common Name	Capt.	M/F	Reference
AMPHIBIA					
Ambystomatidae	Ambystoma macrodactylum krausei	Eastern long-toed salamander	5.1	x	[2]
	Ambystoma macrodactylum sigillatum	Southern long-toed salamander	5.1	x	[2]
	Ambystoma maculatum	Spotted salamander	25.0	x	[7]
	Ambystoma opacum	Marbled salamander	3.5	x	[2]
	Ambystoma opacum	Marbled salamander	4.0	x	[8]
	Ambystoma ordinarium	Puerto Hondo stream salamander	2.0	x	[2]
	Ambystoma talpoideum	Mole salamander	2.3	x	[2]
	Ambystoma texanum	Smallmouth salamander	5.3	x	[2]
	Ambystoma tigrinum	Tiger salamander	10.3	x	[2]
	Ambystoma tigrinum	Tiger salamander	25.0	x	[8]
	Rhyacotriton olympicus olympicus	Olympic salamander	2.7	m	[2]
Amphiumidae	Amphiuma means	Two-toed amphium	14.6	x	[2]
	Amphiuma means	Two-toed amphium	27.0	x	[7]
	Amphiuma tridactylum	Three-toed amphium	12.3	x	[2]
Atelopodidae	Atelopus varius zeteki	Veragoa stubfoot toad	0.8	x	[2]
Bufonidae	Bufo alvarius	Colorado river toad	9.2	x	[2]
	Bufo americanus	American toad	4.7	x	[2]
	Bufo americanus	American toad	30.0	x	[8]
	Bufo americanus	American toad	5.0	x	[10]
	Bufo blombergi	Colombian giant toad	10.8	m	[2]
	Bufo boreas	Western toad	6.0	x	[2]
	Bufo bufo	Common European toad	36.0	x	[7]
	Bufo debilis insidior	Green toad	3.2	x	[2]
	Bufo luetkeni	Yellow toad	9.9	x	[2]
	Bufo marinus	Marine toad	8.0	f	[2]
	Bufo mazatlanensis	Sinaloa toad	3.8	x	[2]
	Bufo paracnemis	Eurura toad	10.3	x	[2]
	Bufo peltacephus		12.7	x	[2]
	Bufo punctatus	Red-spotted toad	4.1	m	[2]
	Bufo quercicus	Oak toad	1.9	x	[2]
	Bufo retiformis	Sonoran green toad	2.9	x	[2]
	Bufo speciosus	Texas toad	3.9	x	[2]
	Bufo viridis	European green toad	3.0	x	[2]
	Bufo woodhousei fowleri	Woodhouse toad	2.4	x	[2]
Caecilidae	Dermorphis mexicanus mexicanus	Mexican caecilian	2.1	f	[2]
Cryptobranchidae	Andrias diavidianus	Chinese giant salamander	4.2	x	[2]
	Andrias japonicus	Japanese giant salamander	16.8	x	[2]
	Andrias japonicus	Japanese giant salamander	55.0	x	[7]
	Cryptobranchus alleganiensis	Eastern hellbender	6.0	x	[2]
	Cryptobranchus alleganiensis	Eastern hellbender	25.0	x	[10]

Dendrobatidae	Dendrobates auratus	Green and black poison frog	8.3	x	[2]
	Dendrobates pumilio	Strawberry poison frog	1.8	x	[2]
Discoglossidae	Bombina bombina	Firebelly toad	5.8	m	[2]
	Bombina bombina	Firebelly toad	20.0	x	[7]
	Bombina orientalis	Oriental firebelly toad	2.3	x	[2]
Hylidae	Agalychnis litodryas	Pink-sided leaf frog	3.5	x	[2]
	Diaglena spatula		7.3	x	[2]
	Hyla avivoca	Bird-voiced treefrog	2.5	x	[2]
	Hyla chrysoscelis	Gray treefrog	2.5	m	[2]
	Hyla cinerea	Green treefrog	6.2	x	[2]
	Hyla crepitans	Emerald-eyed treefrog	4.9	x	[2]
	Hyla femoralis	Pine woods treefrog	2.5	m	[2]
	Hyla gabbi		2.3	x	[2]
	Hyla gratiosa	Barking treefrog	7.2	x	[2]
	Hyla microcephala	Yellow treefrog	2.9	x	[2]
	Hyla pellucens	Palmar treefrog	3.5	x	[2]
	Hyla quinquefasciata		3.5	x	[2]
	Hyla rosenbergi	Rosenberg's treefrog	3.5	m	[2]
	Hyla septentrionalis		12.9	f	[2]
	Hyla vasta	Hispaniola treefrog	5.1	m	[2]
	Pachymedusa dacnicolor	Mexican leaf frog	7.5	f	[2]
	Phrynohyas hebes		5.7	x	[2]
	Pternohyla fodiens	Northern casque-headed frog	5.1	x	[2]
	Smilisca baudini baudini	Mexican treefrog	5.2	x	[2]
	Smilisca phaeota		3.5	x	[2]
	Trachycephalus jordani	Jordan's casque-headed treefrog	3.5	x	[2]
	Triprion spatulatus reticulatus	Shovel-headed treefrog	8.8	x	[2]
Hynobiidae	Hynobius boulengeri	Boulenger's oriental salamander	5.3	m	[2]
Leptodactylidae	Ceratophrys calcarata	Colombian horned frog	10.8	m	[2]
	Ceratophrys cornuta	Surinam horned frog	10.3	f	[2]
	Ceratophrys ornata	Ornate horned frog	11.3	f	[2]
	Leptodactylus pentadactylus	Tungara frog	15.7	x	[2]
	Leptodactylus pentadactylus	S. American bullfrog	12.0	x	[7]
	Physalaemus pustulosus		7.4	x	[2]
Microhylidae	Gastrophryne carolinensis	Eastern narrow-mouthed toad	6.1	x	[2]
	Gastrophryne carolinensis	Eastern narrow-mouthed toad	6.0	x	[7]
	Kaloula pulchra	Malaysian narrowmouth toad	6.0	x	[7]
	Koula pulchra		7.9	x	[2]
Necturidae	Necturus punctatus	Dwarf waterdog	4.5	x	[2]
Pelobatidae	Pelobates fuscus	Common Eurasian spadefoot toad	11.0	x	[7]
	Scaphiopus couchi	Couch spadefoot toad	3.3	x	[2]
	Scaphiopus holbrooki holbrooki	Eastern spadefoot toad	12.3	x	[2]
Pelodryadidae	Litoria caerulea	White's treefrog	8.5	x	[2]
	Litoria caerulea	White's treefrog	16.0	x	[7]
Pipidae	Pipa pipa	Surinam toad	6.8	x	[2]
	Xenopus laevis	African clawed frog	8.8	x	[2]
	Xenopus laevis	African clawed frog	15.0	x	[7]
Plethodontidae	Bolitoglossa subpalmata	La Palma salamander	1.9	x	[2]
	Desmognathus aeneus	Seepage salamander	4.0	x	[2]
	Desmognathus auriculatus	Southern dusky salamander	3.8	x	[2]

	Desmognathus ochrophaeus	Mountain dusky salamander	5.3	x	[2]
	Desmognathus quandramaculatus	Blackbelly salamander	10.3	x	[10]
	Desmognathus welteri	Black mountain salamander	5.4	x	[2]
	Ensatina eschscholtzi croceater	Yellow-blotched salamander	3.3	m	[2]
	Ensatina eschscholtzi klauberi	Large-blotched salamander	4.3	x	[2]
	Ensatina eschscholtzi oregonensis	Oregon salamander	4.5	x	[2]
	Eurycea longicauda guttolineata	Three-lined salamander	5.0	x	[2]
	Eurycea longicauda longicauda	Long-tailed salamander	4.7	x	[2]
	Eurycea wilderae	Blue Ridge two-lined salamander	8.0	x	[10]
	Gyrinophilus palleucus	Tennessee cave salamander	3.9	x	[2]
	Gyrinophilus porphyriticus danielsi	Blue Ridge spring salamander	2.9	x	[2]
	Gyrinophilus porphyriticus porphyriticus	Northern spring salamander	4.8	x	[2]
	Haideotriton wallacei	Georgia blind salamander	1.7	x	[2]
	Hemidactylium scutatum	Four-toed salamander	5.5	f	[2]
	Phaeognathus hubrichti	Red Hills salamander	5.1	x	[2]
	Plethodon elongatus	Del Norte salamander	4.5	x	[2]
	Plethodon glutinosus	Northern slimy salamander	5.5	x	[2]
	Plethodon jordani	Jordan's salamander	5.3	x	[2]
	Plethodon longicrus		5.2	x	[2]
	Plethodon vehiculum	Western red-backed salamander	4.6	x	[2]
	Pseudotriton montanus diastictus	Midland mud salamander	5.5	x	[2]
	Pseudotriton ruber ruber	Northern red salamander	5.5	m	[2]
	Pseudotriton ruber schencki	Blackchin red salamander	5.4	m	[2]
	Pseudotriton ruber vioscai	Southern red salamander	2.5	x	[2]
	Typhloriton spelaeus	Grotto salamander	5.3	x	[2]
Proteidae	Proteus anquinus	Olm	3.1	m	[2]
	Proteus anquinus	Olm	15.0	x	[7]
Ranidae	Mantella aurantiaca	Golden frog	3.7	f	[2]
	Rana adspera		8.6	x	[2]
	Rana catesbeiana	Bullfrog	6.6	m	[2]
	Rana catesbeiana	Bullfrog	16.0	x	[7]
	Rana catesbeiana	Bullfrog	8.0	x	[10]
	Rana occipitalis		10.9	x	[2]
Salamandridae	Cynops pyrrhogaster	Japanese firebelly newt	25.0	x	[7]
	Notophthalmus perstriatus	Striped newt	12.9	m	[2]
	Notophthalmus viridescens	Eastern or red-spotted newt	15.0	x	[8]
	Notophthalmus viridescens viridescens	Eastern or red-spotted newt	3.1	x	[2]
	Pleurodeles walti	Spanish ribbed newt	10.3	x	[2]
	Salamandra salamandra terrestris	European fire salamander	6.0	m	[2]
	Triturus alpestris	Laureutis alpine salamander	3.1	x	[2]
	Triturus cristatus	Northern crested newt	4.0	x	[2]
	Triturus pyrrhogaster		4.4	x	[2]
	Trylototriton verrucosus	Crocodile newt	5.3	x	[2]

Sirenidae	Siren intermedia	Eastern lesser siren	6.3	x	[2]
	Siren lacertina	Greater siren	14.8	f	[2]
	Siren lacertina	Greater siren	25.0	x	[7]
Typholnectidae	Typhlonectes compressicauda	Cayenne caecilian	4.9	x	[2]
REPTILIA					
Acrochordidae	Acrochordus javanicus	Java file snake	5.8	x	[2]
Agamidae	Agama colonorum	West African spinous lizard	1.3	x	[5]
	Agama hartmanni	Hartmann's Sudan lizard	1.2	x	[5]
	Agama mutabilis	Egyptian changeable lizard	2.8	x	[5]
	Agama pallida	Egyptian pale lizard	1.8	x	[5]
	Agama savignyi	Savigny's Egyptian judge-of-the-desert	2.5	x	[5]
	Agama stellio		2.3	f	[2]
	Agama stellio	Levant starred lizard	4.1	x	[5]
	Amphibolurus barbatus		9.9	m	[2]
	Amphibolurus nobbi	Nobbi	2.2	f	[2]
	Calotes emma	East Indian emma lizard	0.7	x	[5]
	Calotes versicolor	Indian changeable lizard	1.5	x	[5]
	Chlamydosaurus kingi	Filled dragon	6.3	x	[2]
	Hydrosaurus amboinensis		3.5	m	[2]
	Leiolepis belliana	Butterfly lizard	5.6	f	[2]
	Lyriocephalus scutatus	Ceylonese knob-nosed lizard	3.5	x	[5]
	Physignathus cocincinus	Water dragon	2.8	f	[2]
	Physignathus leusueri	Australian water lizard	4.1	x	[5]
	Physignathus leusueri	Eastern water dragon	7.5	x	[2]
	Physignathus leusueri	Eastern water dragon	6.0	x	[7]
	Uromastix acanthinurus	N. African spiny dabb lizard	11.4	x	[5]
	Uromastix aegyptius	Egyptian dabb lizard	11.4	x	[5]
Alligatoridae	Alligator mississippiensis	American alligator	47.8	x	[2]
	Alligator mississippiensis	American alligator	56.0	x	[7]
	Alligator mississippiensis	American alligator	34.8	f	[5]
	Alligator sinensis	Chinese alligator	27.1	x	[5]
	Alligator sinensis	Chinese alligator	38.1	m/f	[2]
	Caimen crocodilus	Narrow-snouted spectacled caiman	21.9	x	[2]
	Caimen latirostris	Broad-nosed caiman	22.0	f	[2]
	Caiman latirostris	S. American broad-snouted caiman	8.3	x	[5]
	Caiman palpebrosus	Brazilian musky caiman	5.8	x	[5]
	Caiman sclerops	American rough eyed caiman	3.4	x	[5]
	Melanosuchus niger	Black caiman	13.1	m	[2]
	Paleosuchus palpebrosus	Dwarf caiman	2.4	f	[2]
	Paleosuchus trigonatus	Smooth-fronted caiman	16.3	x	[2]
Amphisbaenidae	Amphisbaena alba	Red worm lizard	13.3	f	[2]
	Amphisbaena alba	American white amphisbaena	1.8	x	[5]
	Amphisbaena braziliana	Brazilian amphisbaena	1.2	x	[5]
	Amphisbaena darwini	Darwin's S. American amphisbaena	0.2	x	[5]
	Blanus cinereus	Spanish grey amphisbaena	0.7	x	[5]
	Lepidosternon microcephalum	Brazilian small-headed amphisbaena	0.2	x	[5]
	Lepidosternon scutigerum	Brazilian shield-bearing amphisbaena	0.9	x	[5]
Anguidae	Anguis fragilis	Slowworm	3.0	x	[5]
	Anguis fragilis	Slowworm	8.3	x	[2]
	Anguis fragilis	Slowworm	54.0	x	[3] [7]

	Gerrhonotus caeruleus	American blue lizard	2.3	x	[5]
	Gerrhonotus liocephus infernalis	Texas alligator lizard	3.5	x	[2]
	Gerrhonotus multicarinatus	Southern alligator lizard	5.2	x	[2]
	Ophisaurus apodus	Armored glass lizard	17.0	x	[2]
	Ophisaurus apus	Palaearctic Scheltopusik limbless lizard or glass snake	10.7	x	[5]
	Ophisaurus koellikeri	Koeliker's glass lizard	9.3	x	[2]
	Ophisaurus ventralis	Eastern glass lizard	3.8	x	[2]
Bipedidae	Bipes biporus	Mole lizard	3.3	x	[2]
Boidae	Acrantophis	Madagascar boa	8.6	m	[2]
	Aspidites melanocephalus	Black-headed python	7.2	x	[2]
	Boa constrictor amarali	Boa constrictor	15.6	x	[2]
	Boa constrictor constrictor	Common boa constrictor	38.8	f	[2]
	Boa constrictor	American boa constrictor	13.3	x	[5]
	Boa constrictor imperator	Albino boa constrictor	29.1	f	[2]
	Boa constrictor ortoni	Boa constrictor	11.9	m	[2]
	Boa madagascariensis	Madagascar boa	19.4	x	[5]
	Calabaria reinhardti	West African ground-python	0.5	x	[5]
	Candoia aspera	New Guinea viper boa	5.9	f	[2]
	Charina bottae bottae	Rubber boa	11.4	x	[2]
	Chondropython viridis	Green tree python	15.1	f	[2]
	Corallus annulatus	Annulated boa	12.3	x	[2]
	Corallus caninus	Emerald tree-boa	15.4	f	[2]
	Corallus enydris cooki	Amazon tree-boa	12.3	f	[2]
	Corallus enydris enydris	Amazon tree-boa	15.4	x	[2]
	Corallus cookii	Cook's American tree-boa	3.3	x	[5]
	Corallus madagascariensis	Madagascar tree-boa (+)	11.5	x	[5]
	Epicrates angulifer	Cuban boa	14.3	f	[2]
	Epicrates angulifer	Cuban pale-headed tree-boa	8.6	x	[5]
	Epicrates cenchria cenchria	Common rainbow boa	10.3	x	[2]
	Epicrates cenchria crassus	Rainbow boa	16.9	f	[2]
	Epicrates cenchria maurus	Rainbow boa	27.3	x	[2]
	Epicrates cenchria	American thick-necked tree-boa	5.5	x	[5]
	Epicrates inornatus	West Indian yellow tree-boa	14.4	x	[5]
	Epicrates inornatus fordi	Puerto Rican boa	7.5	m	[2]
	Epicrates striatus fosteri	Haitian boa	10.9	f	[2]
	Epicratus striatus ssp.	Haitian boa	22.1	m	[2]
	Epicratus subflavus	Jamaica boa	9.8	x	[2]
	Eryx colubrinus	Kenya sand-boa	10.4	x	[2]
	Eryx conicus	Rough-scaled sand-boa	14.4	f	[2]
	Eryx jaculus	Caucasian sand-boa	18.3	f	[2]
	Eryx jaculus	Egyptian sand-boa (+)	13.5	x	[5]
	Eryx johnii	John's Indian sand-boa (+)	13.0	x	[5]
	Eryx johnii	John's sand-boa	19.5	x	[2]
	Eryx muelleri	Mueller's N. African sand-boa	2.5	x	[5]
	Eryx tataricus	Tartar sand-boa	8.3	x	[2]
	Eryx thebaibus	Theban sand-boa	3.1	x	[5]
	Eunectes barbouri	Barbour's anaconda	19.5	f	[2]
	Eunectes deschauenseei	Dark-spotted anaconda	8.3	f	[2]
	Eunectes murinus	American anaconda	8.0	x	[5]

	Eunectes murinus	Green anaconda	13.9	x	[2]
	Eunectes murinus	Green anaconda	29.0	x	[7]
	Eunectes notaeus	Paraguayan anaconda	3.4	x	[5]
	Liasis amethystinus kinghorni		13.8	x	[2]
	Liasis childreni		9.8	x	[2]
	Liasis fuscus fuscus	Common brown water python	10.2	x	[2]
	Liasis fuscus ssp.	Brown water python	18.7	f	[2]
	Liasis mackloti mackloti	Macklot's python	6.3	x	[2]
	Lichanura trivirgata roseofusca	Coastal rosy boa	18.6	x	[2]
	Loxocemus bicolor	Mexican burrowing python	32.8	m	[2]
	Morelia argus argus		9.3	f	[2]
	Morelia argus variegata		12.8	m	[2]
	Python curtus curtus	Short-tailed python	27.8	f	[2]
	Python molurus bivittatus	Albino or heterozygous Burmese python	28.3	f	[2]
	Python molurus molururs	Indian python	34.2	m	[2]
	Python molurus	Indian python or rock-python	15.8	x	[5]
	Python regius	African royal python	8.7	x	[5]
	Python regius	Ball python	30.5	m	[2]
	Python reticulatus	Reticulate python	25.3	m	[2]
	Python reticulatus	East Indian reticulated python	21.3	x	[5]
	Python sebae	African python	27.3	f	[2]
	Python sebae	African python	18.0	x	[5]
	Python spilotes	Australian diamond-python	4.3	x	[5]
	Sanzinia madagascariensis	Madagascar tree-boa	8.0	x	[2]
	Tropidophis canus curtus	Great Inagua Island dwarf boa	4.0	m	[2]
	Tropidophis caymanensis parkeri	Cayman Islands dwarf boa	8.7	x	[2]
	Ungaliophis continentalis	Isthmian dwarf boa	17.7	x	[2]
Carettochelyidae	Carettochelys insculpta	Pignose turtle	17.3	f	[2]
Chamaeleonidae	Chamaeleo basiliscus	N. African basilisk chameleon	1.8	x	[5]
	Chamaeleo pumilus		3.4	x	[5]
	Chamaeleo chamaeleon	Common chameleon	1.0	f	[2]
	Chamaeleo jacksoni	Jacson's chameleon	1.0	f	[2]
	Chamaeleo melleri	Meller's chameleon	1.0	f	[2]
Chelidae	Batrachemys dahli		4.9	m	[2]
	Batrachemys nasuta		13.5	f	[2]
	Chelodina longicollis	Common snakeneck turtle	35.1	x	[2]
	Chelodina longicollis	Common snakeneck turtle	37.0	x	[7]
	Chelodina oblonga	Narrow-breasted snakeneck turtle	4.4	m	[2]
	Chelodina longicollis	Australian long-necked terrapin	7.2	x	[5]
	Chelus fimbriata	S. American matamata terrapin	6.0	x	[5]
	Chelus fimbriata		16.2	m	[2]
	Emydura latisternum	Australian broad-breasted terrapin	5.8	x	[5]
	Hydromedusa maximiliani	Maximilian's Brazilian terrapin	11.8	x	[5]
	Hydromedusa tectifera	S. American snakeneck turtle	8.8	f	[2]
	Mesoclemmys gibba		6.3	x	[2]
	Phrynops geoffroanus hilairi	Geoffrey's sideneck turtle	34.8	f	[2]
	Platemys platycephala	Twistneck turtle	19.7	f	[2]
	Platemys spixii	Spix's Brazilian terrapin	8.2	x	[5]
Cheloniidae	Caretta caretta	Loggerhead turtle	33.0	x	[7]
	Lepidochelys kempi	Atlantic ridley	3.6	f	[2]

	Hydraspis geoffroyana	Geoffroy's Brazilian terrapin	7.5	x	[5]
	Hydraspis hilarii	St. Hilaire's S. American terrapin	13.8	x	[5]
Chelydridae	Chelydra serpentina	N. American alligator terrapin	17.8	x	[5]
	Chelydra serpentina rossignoni	Snapping turtle	23.2	x	[2]
	Chelydra serpentina serpentina	Common snapping turtle	38.7	x	[2]
	Chelydra serpentina serpentina	Common snapping turtle	47.0	x	[8]
	Macroclemys temminckii	N. American snapping turtle	5.3	x	[5]
	Macroclemys temmincki	Alligator snapping turtle	58.8	m	[2]
	Macroclemys temmincki	Alligator snapping turtle	59.0	x	[7]
Cinosternidae	Cinosternum cruentatum	Central American gory mud terrapin	6.8	x	[5]
	Cinosternum leucostomum	American white-mouthed mud terrapin	17.7	x	[5]
	Cinosternum odoratum	N. American stink-pot mud terrapin	22.6	x	[5]
	Cinosternum pennsylvanicum	Pennsylvania mud terrapin	7.1	x	[5]
	Cinosternum scorpioides	S. American scorpion mud terrapin	8.2	x	[5]
Colubridae	Ablabophis rufulus	S. African rufous snake	4.3	x	[5]
	Acrochordus javanicus	E. Indian water-elephant snake	4.1	x	[5]
	Ahaeteulla tristis		4.3	f	[2]
	Alsophis portoricensis	Puerto Rican pacer	2.0	x	[2]
	Arizona elegans noctivaga	Glossy snake	12.2	x	[2]
	Arizona elegans occidentalis	California glossy snake	10.1	m	[2]
	Boaedon fuliginosus		7.0	f	[2]
	Boiga blandingi	Blanding's cat snake	8.8	f	[2]
	Boiga cynodon	Dog-tooth cat snake	9.1	m	[2]
	Boiga dendrophila	Mangrove snake	13.0	x	[2]
	Boiga irregularis	Brown tree snake	9.3	m	[2]
	Boiga trigonata	Indian gamma snake	8.2	x	[2]
	Boodon lineatus	African lineated snake	3.6	x	[5]
	Chilomeniscus cinctus	Banded sand snake	4.0	x	[2]
	Chrysopelea ornata	Golden treesnake	4.3	x	[2]
	Clelia clelia	Mussarana	11.5	m	[2]
	Coelopeltis moilensis	Moila snake	0.5	x	[5]
	Coelopeltis monspessulana	Paearctic montpellier	1.8	x	[5]
	Coluber corais couperi	American corais snake	3.9	x	[5]
	Coluber guttatus	N. American corn-snake	3.3	x	[5]
	Coluber longissimus	European aesculapian snake	3.9	x	[5]
	Coluber quatuorlineatus	European four-lined snake	4.6	x	[5]
	Conophis vittatus vittatus	Striped road guarder	2.1	x	[2]
	Coronella austriaca	European smooth snake	1.3	x	[5]
	Coronella getula	N. American king snake	3.8	x	[5]
	Crotaphopeltis hotamboeia	Herald snake	2.3	f	[2]
	Cyclagras gigas		16.8	f	[2]
	Dasypeltis atra	African egg-eating snake	5.2	f	[2]
	Dasypeltis medici	East African egg-eating snake	4.0	f	[2]
	Dasypeltis scabra	African egg-eating snake	13.6	x	[2]
	Dasypeltis scabra	African rough-keeled snake	3.8	x	[5]
	Dendraspis viridis	Tropical African green treesnake	3.8	x	[5]
	Dendrophis punctulatus	Australian punctulated treesnake	3.1	x	[5]
	Dinodon rufozonatum	Red barded snake	13.7	f	[2]
	Dispholidus typhus	Boomslang	8.7	m	[2]
	Drymarchon corais corais	Common indigo snake	5.1	x	[2]
	Drymarchon corais couperi	Eastern indigo snake	25.9	x	[2]

Drymarchon corais couperi	Eastern indigo snake	25.0	x	[7]
Drymarchon corais melanurus	Indigo snake	9.0	f	[2]
Drymarchon corais rubidus	Indigo snake	11.6	x	[2]
Drymobius margaritiferus margaritiferus	Speckled racer	4.3	f	[2]
Dryophis mycterizans	Indian long-snouted treesnake	2.4	x	[5]
Elaphe climacophora	Japanese rat snake	12.0	m	[2]
Elaphe flavirufa	Yellow-red rat snake	7.8	x	[2]
Elaphe guttata emoryi	Great plains rat snake	21.1	x	[2]
Elaphe guttata guttata	Corn snake	21.8	x	[2]
Elaphe longissima	Aesculapian rat snake	3.2	x	[2]
Elaphe obsoleta bairdi	Rat snake	9.4	x	[2]
Elaphe obsoleta lindheimeri	Texas rat snake	7.0	x	[2]
Elaphe obsoleta obsoleta	Black rat snake	20.1	x	[2]
Elaphe obsoleta quadrivittata	Yellow rat snake	17.1	m	[2]
Elaphe obsoleta spiloides	Gray rat snake	13.9	x	[2]
Elaphe oxycephala		5.4	f	[2]
Elaphe quadrivirgata	Japanese four-lined rat snake	16.8	f	[2]
Elaphe quatuorlineata	Four-lined rat snake	5.2	x	[2]
Elaphe situla	European leopard snake	23.0	x	[7]
Elaphe subocularis	Trans-pecos rat snake	13.8	x	[2]
Elaphe taeniura	Taipan beauty snake	9.7	x	[2]
Elaphe triaspis intermedia	Western green rat snake	9.5	f	[2]
Elaphe vulpina gloydi	Eastern fox snake	7.4	f	[2]
Elaps lemniscatus	S. American checkered snake	10.0	x	[5]
Enhydris chinensis	Chinese water snake	4.2	f	[2]
Farancia abacura reinwardti	Western mud snake	18.0	f	[2]
Helicops angulatus	S. American angulated snake	3.0	x	[5]
Herpetodryas carinatus	American carinated snake	5.4	x	[5]
Herpeton tentaculatum	East Indian tentacled snake	2.9	x	[5]
Heterodon nasicus	N. American long-nosed snake	5.6	x	[5]
Heterodon nasicus kennerlyi	Mexican hognose snake	9.2	x	[2]
Heterodon nasicus nasicus	Plains hognose snake	8.2	m	[2]
Heterodon platyrhinos	Eastern hognose snake	4.0	m	[2]
Heterodon platyrhinos	N. American hognose snake	3.5	x	[5]
Hypsiglena torquata texana	Night snake	9.3	x	[2]
Lampropeltis calligaster calligaster	Prairie kingsnake	11.0	f	[2]
Lampropeltis getulus californiae	California kingsnake	14.7	m	[2]
Lampropeltis getulus floridana	Florida kingsnake	12.0	m	[2]
Lampropeltis getulus getulus	Common kingsnake	21.4	m	[2]
Lampropeltis getulus holbrooki	Speckled kingsnake	14.4	m	[2]
Lampropeltis getulus niger	Black kingsnake	13.4	m	[2]
Lampropeltis getulus nigritus	Black desert kingsnake	11.1	m	[2]
Lampropeltis getulus splendida	Desert kingsnake	12.8	f	[2]
Lampropeltis getulus yumensis		23.3	x	[2]
Lampropeltis pyromelana infrabialis	Utah mountain kingsnake	4.4	x	[2]
Lampropeltis pyromelana pyromelana	Arizona mountain kingsnake	15.3	f	[2]

Lampropeltis pyromelana woodini	Huachuca mountain kingsnake	9.3	m	[2]
Lampropeltis triangulum amaura	Louisiana milk snake	20.6	x	[2]
Lampropeltis triangulum annulata	Mexican milk snake	20.2	x	[2]
Lampropeltis triangulum elapsoides	Scarlet kingsnake	13.0	f	[2]
Lampropeltis triangulum syspila	Red milk snake	10.0	x	[2]
Lampropeltis triangulum triangulum	Milk snake	21.3	x	[2]
Lampropeltis zonata multicincta	Sierra mountain kingsnake	16.3	m	[2]
Lampropeltis zonata multifasciata	Coast mountain kingsnake	13.4	x	[2]
Lampropeltis zonata parvirubra	San Bernadino mountain kingsnake	11.9	x	[2]
Lampropeltis zonata zonata	Saint Helena mountain kingsnake	21.2	m	[2]
Leptodeira annulata	Banded cat-eyed snake	3.4	x	[2]
Leptodeira annulata	American annulated snake	4.9	x	[5]
Leptodeira hotemboeia	African rufescent snake	4.3	x	[5]
Leptophis mexicanus	Mexican parrot snake	7.7	x	[2]
Lioheterodon madagascariensis	Madagascar sharp-nosed snake	3.4	x	[5]
Liophis anomala	Strange ground snake	4.8	x	[2]
Lytorhynchus diadema	Diademed sand-snake	0.4	x	[5]
Macroprotodon cucullatus	Mediteranean hooded snake	5.9	x	[5]
Masticophis flagellum flagellum	Coachwhip	16.6	m	[2]
Masticophis flagellum flavigularis		13.4	x	[2]
Masticophis flagellum piceus	Red coachwhip	12.1	m	[2]
Masticophis flagellum testaceus	Central coachwhip	8.3	x	[2]
Mehelya capensis	Lapi file snake	10.9	x	[2]
Naia bungarus	Asiatic hamadryad	12.6	x	[5]
Naia flava	African yellow cobra	6.0	x	[5]
Naia haie	Egyptian cobra	6.3	x	[5]
Naia melanoleuca	African black and white cobra	3.8	x	[5]
Naia tripudians	Indian cobra (+)	12.0	x	[5]
Natrix fasciata	N. American mocassin snake	9.4	x	[5]
Natrix fasciata rhombifer	N. American rhombmarked snake	7.6	x	[5]
Natrix natrix	Grass snake	9.0	x	[4]
Natrix natrix	Palaearctic grass-snake	7.4	x	[5]
Natrix piscator	Indian river-snake	9.0	x	[5]
Natrix saurita	N. American ribbon-snake	6.6	x	[5]
Natrix septemvittata	N. American seven-banded snake	5.5	x	[5]
Natrix sirtalis	N. American striped snake	5.8	x	[5]
Natrix tessellata	Dice snake	2.3	x	[2]
Natrix tessellata	Palaearctic tessellated snake	2.7	x	[5]
Natrix tigrina tigrina		3.1	x	[2]
Natrix viperina	European viperine snake	9.2	x	[5]

Oxybelis aeneus	Brown vine snake	5.0	x	[2]
Petalognathus nebulatus	S. American clouded snake	4.3	x	[5]
Peudaspis cana	S. African hoary snake	6.2	x	[5]
Philodryas viridissimus	S. American all-green treesnake	4.1	x	[5]
Pituophis melanoleucus	N. American pine-snake	4.5	x	[5]
Pituophis melanoleucus affinis		15.0	f	[2]
Pituophis melanoleucus annectens	San Diego gopher snake	20.4	m	[2]
Pituophis melanoleucus catenifer	Pacific gopher snake	11.0	x	[2]
Pituophis melanoleucus deserticola	Great Basin gopher snake	15.8	f	[2]
Pituophis melanoleucus melanoleucus	Gopher snake	20.8	f	[2]
Pituophis melanoleucus mugitus	Florida pine snake	13.0	x	[2]
Pituophis melanoleucus sayi	Bull snake	22.4	x	[2]
Pituophis sayi	N. American bull snake	3.0	x	[5]
Psammophis elegans	West African slender snake	1.9	x	[5]
Psammophis schokari	Schokari snake	0.9	x	[5]
Psammophis sibilans	African hissing snake	10.3	x	[5]
Psammophis subtaeniatus	Stripe-belly sand racer	5.8	x	[2]
Pseudaspis cana	Mole snake	8.3	m	[2]
Pseudechis porphyriacus	Australian purplish death adder	5.8	x	[5]
Ptyas mucosus	Oriental rat snake	11.1	m	[2]
Rhadinaea merremii	Merrem's snake	4.3	x	[5]
Rhamphiophis multimaculata	S. African many-spotted snake	10.0	x	[5]
Rhamphiophis oxyrhychus	Rufous beaked snake	13.3	m	[2]
Rhinocheilus lecontei lecontei	Western long-nosed snake	18.3	f	[2]
Rhinocheilus lecontei tessellatus	Texas long-nosed snake	16.4	m	[2]
Salvadora hexalepis hexalepis	Western patch-nosed snake	14.3	m	[2]
Salvadora mexicana	Mexican patch-nosed snake	6.3	x	[2]
Sepedon haemachates	S. African ring-hals snake	4.6	x	[5]
Spalerosophis diadema	Diademed snake	9.6	f	[2]
Spilotes pullatus	Tropical rat snake	13.6	f	[2]
Storeria occipitomaculata occipitomaculata	Red-bellied snake	2.2	x	[2]
Tarbophis obtusus	African fierce-eyed snake	2.4	x	[5]
Telescopus semiannulatus	Eastern tiger snake	7.6	x	[2]
Thamnophis butleri	Butler's garter snake	2.0	x	[2]
Thamnophis couchi hammondi	Western aquatic garter snake	7.7	x	[2]
Thamnophis cyrtopsis ocellata	Eastern blackneck garter snake	6.0	m	[2]
Thamnophis elegans	Mountain garter snake	6.1	x	[2]
Thamnophis marcianus	Checkered garter snake	7.2	f	[2]
Thamnophis proximus proximus	Western ribbon snake	3.6	x	[2]
Thamnophis radix hayden	Western plains garter snake	3.6	m	[2]
Thamnophis sauritus sauritus	Eastern ribbon snake	3.9	x	[2]
Thamnophis sirtalis sirtalis	Common garter snake	10.0	x	[2]
Thrasops jacksoni	Jackson's black treesnake	5.8	x	[2]

	Tretanorhinus variabilis lewisi	Variable swamp snake	4.3	f	[2]
	Trimorphodon fasciolata		4.8	x	[2]
	Trimorphodon lambda	Sonoran lyre snake	5.3	x	[2]
	Trimorphodon vandenburghi	California lyre snake	6.0	x	[2]
	Uromacer oxyrhynchus	Painted snake	12.1	m/f	[2]
	Walterinnesia aegyptia	Innes Bey's Egyptian black snake	0.7	x	[5]
	Zamenis diadema	Clifford's snake (+)	13.6	x	[5]
	Zamenis florulentus	Egyptian flowered snake	9.3	x	[5]
	Zamenis hippocrepis	European horseshoe-snake	6.3	x	[5]
	Zamenis mucosus	Indian rat snake	11.3	x	[5]
	Zamenis ravergieri nummifer	Palaearctic tail-lined snake (+)	13.8	x	[5]
	Zamenis rogersi	Roger's Egyptian snake	5.7	x	[5]
	Zamenis ventrimaculatus	West Asian spot-bellied snake	5.8	x	[5]
	Zamensis gemonensis asianus	Syrian black snake	10.3	x	[5]
Cordylidae	Cordylus giganteus	Giant spinytail lizard	5.3	x	[2]
	Cordylus giganteus	Giant spinytail lizard	5.0	x	[7]
	Cordylus polyzonus	African spinytail lizard	9.1	x	[2]
	Cordylus warreni	Warren's spinytail lizard	4.3	x	[2]
	Gerrhosaurus major	Rough-scaled plated lizard	11.1	f	[2]
	Gerrhosaurus nigrolineatus auritus	Black-lined plated lizard	8.3	x	[2]
	Gerrhosaurus validus	Giant plated lizard	4.0	x	[2]
	Platysaurus guttatus	Lesser flat lizard	10.3	m	[2]
	Platysaurus intermedius subniger	Common flat lizard	3.1	f	[2]
	Zonosaurs laticaudatus	Western girdled lizard	8.0	x	[2]
Crocodylidae	Crocodylus porosus	Indian estuarine crocodile	8.8	x	[5]
	Crocodylus americanus	American sharp-nosed crocodile	9.0	x	[5]
	Crocodylus cataphractus	West African long-nosed crocodile	7.8	x	[5]
	Crocodylus niloticus	Nile crocodile	15.0	x	[5]
	Crocodylus rhombifer	Cuban crocodile	12.8	x	[5]
	Crocodylus tetraspis	West African broad-fronted crocodile	15.0	x	[5]
	Crocodylus acutus	American crocodile	32.9	f	[2]
	Crocodylus cataphractus	African slender-snouted crocodile	42.4	m	[2]
	Crocodylus intermedius	Orinoco crocodile	17.3	m	[2]
	Crocodylus moreleti	Morelet's crocodile	25.4	f	[2]
	Crocodylus niloticus	Nilotic crocodile	18.4	m	[2]
	Crocodylus palustris	Mugger or marsh crocodile	28.4	x	[2]
	Crocodylus porosus	Salt-water crocodile	41.7	m	[2]
	Crocodylus rhombifer	Cuban crocodile	18.3	m/f	[2]
	Osteolaemus tetraspis	West African dwarf crocodile	42.2	f	[2]
	Tomistoma schlegeli	False gavial	14.3	f	[2]
Dermatemydidae	Dermatemys mawii	Mawe's American terrapin	8.1	x	[5]
Elapidae	Acanthophis antarcticus antarcticus	Common death adder	7.2	m/f	[2]
	Acanthophis antarcticus laevis	Common death adder	6.2	m	[2]
	Acanthophis antarcticus pyrrhus	Common death adder	3.4	x	[2]
	Acanthophis antarcticus rugosus	Common death adder	6.2	f	[2]
	Aspidelaps scutatus	Shieldnose cobra	11.1	x	[2]

	Bungarus caeruleus	Indian krait	10.5	x	[2]
	Bungarus fasciatus	Red-headed krait	11.5	x	[2]
	Bungarus multicinctus	Many-banded krait	9.4	m	[2]
	Dendroaspis angusticeps	Common mamba	14.3	f	[2]
	Dendroaspis jamesoni	Jameson's mamba	9.3	x	[2]
	Dendroaspis polylepis	Black mamba	10.1	f	[2]
	Dendroaspis viridis	Green mamba	12.5	x	[2]
	Elapsoidea sundevalli	Sundevall's garter snake	8.6	x	[2]
	Hemachatus haemachatus	Ringneck spitting cobra	11.8	x	[2]
	Micrurus affinis affinis		4.4	f	[2]
	Micrurus fulvius	Eastern coral snake	6.8	x	[2]
	Naja haje	Egyptian cobra	9.2	m	[2]
	Naja melanoleuca	Black-lipped cobra	29.1	f	[2]
	Naja melanoleuca	Black-lipped cobra	29.0	x	[7]
	Naja naja atra	Indian cobra	11.7	x	[2]
	Naja naja kaouthia	Indian cobra	12.8	x	[2]
	Naja naja naja	Indian cobra	17.3	m	[2]
	Naja naja samarensis	Indian cobra	11.4	x	[2]
	Naja naja sputatrix	Indian cobra	13.5	x	[2]
	Naja naja ssp.	Indian cobra	23.9	f	[2]
	Naja nigricollis mossambica	Blackneck spitting cobra	13.0	m	[2]
	Naja nigricollis ssp.	Blackneck spitting cobra	22.1	x	[2]
	Naja nivea	Cape cobra	23.6	x	[2]
	Notechis scutatus	Mainland tiger snake	14.1	m	[2]
	Ophiophagus hannah	King cobra	17.1	m	[2]
	Pseudechis australis	King grown snake	11.1	x	[2]
	Pseudechis porphyriacus	Red-belly black snake	5.1	x	[2]
Emydidae	Chinemys megalocephala		7.7	m	[2]
	Chinemys reevesi	Reeve's turtle	24.3	x	[2]
	Clemmys guttata	Spotted turtle	8.3	m	[2]
	Clemmys guttata	Spotted turtle	26.0	x	[8]
	Clemmys guttata	Spotted turtle	6.1	x	[5]
	Clemmys bealei		27.3	x	[2]
	Clemmys caspica rivulata		13.3	m	[2]
	Clemmys insculpta	Wood turtle	12.5	x	[2]
	Clemmys insculpta	Wood turtle	60.0	x	[8]
	Clemmys leprosa	Spanish terrapin	5.1	x	[5]
	Clemmys muhlenbergi	Muhlenberg's turtle	13.6	x	[2]
	Clemmys mutica		7.1	f	[2]
	Deirochelys reticularia	Chicken turtle	6.1	f	[2]
	Emydoidea blandingii	Blanding's turtle	12.8	x	[2]
	Emys blandingii	Blanding's N. American pond tortoise	3.0	x	[5]
	Emys orbicularis	Swamp turtle or tortoise	11.7	m	[2]
	Emys orbicularis	European pond tortoise	27.9	x	[5]
	Geoclemys hamiltoni	Spotted pond turtle	14.3	m	[2]
	Graptemys barbouri	Barbour's map turtle	5.8	x	[2]
	Graptemys geographica	Common map turtle	5.5	x	[2]
	Graptemys kohni	Mississippi map turtle	19.8	m	[2]
	Graptemys oculifera	Ringed map turtle	4.7	f	[2]
	Graptemys pseudogeographica	False map turtle	32.5	x	[2]
	Graptemys pulchra	Alabama map turtle	15.7	x	[2]

	Heiremys annandalii	Yellow-headed temple turtle	9.3	f	[2]
	Maclemys terrapin	Diamondback terrapin	9.3	m	[2]
	Notochelys platynota	Malayan flatshell turtle	8.4	f	[2]
	Ocadia sinensis	Chinese stripeneck turtle	12.4	f	[2]
	Terrapene carolina bauri	Florida box turtle	22.6	x	[2]
	Terrapene carolina carolina	Eastern box turtle	26.4	m	[2]
	Terrapene carolina carolina	Eastern box turtle	75.0	x	[8]
	Terrapene carolina major	Gulf coast box turtle	21.6	x	[2]
	Terrapene carolina triunguis	Three-toed box turtle	24.3	m	[2]
	Terrapene coahuila	Coahuilan box turtle	9.4	f	[2]
	Terrapene mexicana yucatana		15.4	m	[2]
	Terrapene nelson	Spotted box turtle	11.8	m	[2]
Gavialidae	Gavialis gangeticus	Indian gavial	27.8	f	[2]
Gekkonidae	Coleonyx variegatus	Western banded gecko	34.8	m	[2]
	Eublepharis macularius	Leopard gecko	4.9	f	[2]
	Gehyra oceanica	Oceanic dtella	15.5	m	[2]
	Gekko gecko	Tokay gecko	10.5	x	[2]
	Gekko verticillatus	East Indian vericillated gecko	3.1	x	[5]
	Gonatodes albogularis fuscus	Yellowhead gecko	13.0	x	[2]
	Gymnodactylus spyrurus		1.5	m	[2]
	Hemidactylus brookii	Brook's gecko	1.1	x	[5]
	Hemidactylus turcicus	Turkish gecko	2.4	x	[5]
	Hoplodactylus pacificus	New Zealand gecko	1.9	x	[5]
	Hoplodactylus pacificus	Pacific sticky-toed gecko	10.7	x	[2]
	Oedura marmorata	Marbled velvet gecko	2.9	x	[2]
	Oedura robusta	Robust velvet gecko	8.1	x	[2]
	Pachydactylus bibronii	Bibron's thick-toed gecko	4.6	m	[2]
	Phelsuma laticauda	Golddust day gecko	4.0	m	[2]
	Phelsuma lineata	Striped day gecko	3.0	x	[2]
	Phelsuma madagascariensis	Madagascar day gecko	10.0	x	[2]
	Phelsuma madagascariensis	Madagascar gecko	1.3	x	[5]
	Ptychozoon homalocephalum	East Indian fringed gecko	1.1	x	[5]
	Ptyodactylus hasselquisti	Fan-footed gecko	3.4	x	[5]
	Stenodactylus petrii	Flinders Petrie's gecko	0.8	x	[5]
	Stenodactylus stenodactylus	Elegant gecko	0.7	x	[5]
	Tarentola annularis	Egyptian gecko	8.0	x	[5]
	Tarentola delalandii	Delalande's West African gecko	2.4	x	[5]
	Tarentola mauretanica	Moorish gecko	5.6	x	[5]
	Thecadactylus rapicaudus	American gecko	0.9	x	[5]
	Tropiocolotes steudneri	Steudner's pygmy gecko	1.8	x	[5]
	Ptyodactylus hasselquisti	Yellow fan-fingered gecko	3.3	m	[2]
	Sphaerodactylus notatus notatus	Florida reef gecko	1.3	x	[2]
	Tarentola mauritanica	Common wall gecko	4.1	x	[2]
	Tarentola mauritanica	Common wall gecko	7.0	x	[7]
	Teratoscincus microlepis	Small-scaled wonder gecko	3.1	x	[2]
	Teratoscincus scincus	Common wonder gecko	9.3	m	[2]
	Uroplates fimbriatus	Madagascar geckoid lizard	1.3	x	[5]
Glauconiidae	Glauconia cairi	Cairo earth-snake	0.2	x	[5]
Helodermatidae	Heloderma horridum alvarezi	Beaded lizard	27.3	m	[2]
	Heloderma horridum horridum	Beaded lizard	20.2	f	[2]

	Heloderma horridum ssp.	Mexican beaded lizard	22.8	f	[2]
	Heloderma suspectum	Arizona poisonous lizard	8.3	x	[5]
	Heloderma suspectum	Gila monster	20.0	x	[7]
	Heloderma suspectum sp.	Gila monster	27.8	x	[2]
	Heloderma suspectum suspectum	Reticulate gila monster	16.4	m	[2]
Iguanidae	Amblyrhynchus cristatus	Marine iguana	6.4	m	[2]
	Anolis alligator	Barbadian anolis	2.0	x	[5]
	Anolis carolinensis	Carolinan anolis	1.3	x	[5]
	Anolis carolinensis carolinensis	Green anole	7.1	m	[2]
	Anolis equestris	Knight anole	8.3	f	[2]
	Anolis lineatopus	Jamaica anolis	1.0	x	[5]
	Anolis luteogularis	White-throated anole	2.8	m	[2]
	Anolis stejnegeri stejnegeri		2.9	m	[2]
	Basiliscus basiliscus	Common basilisk	3.3	x	[2]
	Basiliscus plumbifrons	Green basilisk	5.1	m	[2]
	Basiliscus vittatus	Brown basilisk	5.8	x	[2]
	Brachylophus fasciatus	Dfigi banded iguana	5.9	m	[2]
	Conolophus cristatus	Smooth helmeted iguana	5.2	x	[2]
	Conolophus pallidus	Santa Fe land iguana	17.1	f	[2]
	Conolophus subcristatus	Galapagos land iguana	7.3	m	[2]
	Conolophus subcristatus	Galapagos land iguana	15.0	x	[7]
	Ctenosaura hemilopha	Cape spinytail iguana	9.6	x	[2]
	Ctenosaura pectinata	Mexican spinytail iguana	8.2	m	[2]
	Ctenosaura similis	Black iguana	4.8	m	[2]
	Cyclura cornuta	Rhinoceros iguana	16.7	m	[2]
	Cyclura figginsi		8.7	x	[2]
	Cyclura inornata		5.6	m	[2]
	Cyclura macleayi caymanensis		8.7	x	[2]
	Cyclura macleayi lewisi		5.6	m	[2]
	Cyclura pinguis	Anegada ground iguana	3.2	f	[2]
	Cyclura rileyi nuchalis	White cay ground iguana	5.6	m	[2]
	Cyclura rileyi ssp.	White cay ground iguana	7.1	x	[2]
	Deiroptyx vermiculatus		1.9	m	[2]
	Dipsosaurus dorsalis	Desert iguana	14.6	m/f	[2]
	Iguana delicatissima	West Indian iguana	4.4	x	[2]
	Iguana delicatissima	American naked-necked iguana	1.0	x	[5]
	Iguana iguana	Green iguana	12.4	f	[2]
	Iguana tuberculata	American tuberculated iguana	2.3	x	[5]
	Iguana tuberculata rhinolopha	American nose crested iguana	1.4	x	[5]
	Leiocephus carinatus	Northern curlytail lizard	10.8	m	[2]
	Mariguana agassizi	White cay ground iguana	5.2	x	[2]
	Metopoceros cornutus	West Indian black iguana	4.2	x	[5]
	Oplurus cyclurus	Merrem's Madagascar swift	9.3	x	[2]
	Phrynosoma douglassi	Short-horned lizard	1.0	x	[2]
	Phrynosoma orbiculare	N. American horned lizard	1.1	x	[5]
	Saauromus obesus tumidus	Arizona chuckwalla	5.5	x	[2]
	Sceloporus cyanogenys		6.7	x	[2]
	Sceloporus woodi	Florida scrub lizard	1.7	m	[2]
	Tropidurus hispidus	Brazilian taraguira lizard	1.6	x	[5]
	Uma notata notata	Colorado desert fringe-toed lizard	8.1	x	[2]

Kinosternidae	Kinosternon baurii	Striped mud turtle	7.6	f	[2]
	Kinosternon cruentatus		6.9	m	[2]
	Kinosternon flavescens	Yellow mud turtle	10.3	x	[2]
	Kinosternon herrerai	Herrera's mud turtle	19.5	m	[2]
	Kinosternon scorpidoides	Scorpian mud turtle	14.0	x	[2]
	Kinosternon sonoriensis	Sonoran mud turtle	27.8	x	[2]
	Kinosternon subrubrum	Common mud turtle	40.0	x	[4]
	Kinosternon subrubrum hippocrepis	Mud turtle	18.3	x	[2]
	Staurotypus salvini	Chiapas giant musk turtle	11.7	m	[2]
	Staurotypus triporcatus	Mexican giant musk turtle	19.8	f	[2]
	Sternotherus carinatus	Keel-backed musk turtle	13.5	x	[2]
	Sternotherus derbianus	Lord Derby's African terrapin	8.4	x	[5]
	Sternotherus minor minor	Loggerhead musk turtle	23.9	m	[2]
	Sternotherus niger	African black terrapin	5.8	x	[5]
	Sternotherus nigricans	Madagascar blackish terrapin	9.2	x	[5]
	Sternotherus odoratus	Stinkpot	54.8	x	[2]
	Sternotherus sinuatus	Natal terrapin	8.2	x	[5]
Lacertidae	Acanthodactylus boskianus	Daudin's fringe-fingered lizard	1.6	x	[5]
	Acanthodactylus pardalis	Leopard fringe-fingered lizard	1.5	x	[5]
	Acanthodactylus scutellatus	Scutellated fringe-fingered lizard	2.9	x	[5]
	Eremias rubropunctata	Red spotted lizard	3.3	x	[5]
	Lacerta agilis	European sand lizard	1.6	x	[5]
	Lacerta dugesi	Madeiran sharp-headed lizard	4.8	x	[5]
	Lacerta galloti	Gallot's Canary Island lizard	5.6	x	[5]
	Lacerta lepida lepida	Jeweled lizard	5.3	x	[2]
	Lacerta muralis	Wall lizard	3.0	x	[4]
	Lacerta muralis filfoliensis	European wall lizard	2.8	x	[5]
	Lacerta ocellata	European eyed lizard	7.1	x	[5]
	Lacerta simonyi	Simony's Canary Island lizard	5.4	x	[5]
	Lacerta viridis	European green lizard	3.2	x	[5]
Pelomedusidae	Pelomedusa galeata	African helmeted terrapin	16.8	x	[5]
	Pelomedusa subrufa olivacea	Helmeted turtle	7.3	f	[2]
	Pelomedusa subrufa subrufa	Helmeted turtle	16.0	x	[2]
	Pelusios adansonii	Adanson's mud turtle	4.7	f	[2]
	Pelusios castaneus	West African mud turtle	41.0	x	[7]
	Pelusios castaneus derbianus	West African mud turtle	24.8	f	[2]
	Pelusios niger	West African black turtle	14.7	f	[2]
	Pelusios sinuatus	East African serrated mud turtle	6.1	f	[2]
	Pelusios subniger	East African black mud turtle	29.3	x	[2]
	Podocnemis sextuberculata	Six-tubercled river turtle	6.1	f	[2]
	Podocnemis cayennensis		5.2	x	[2]
	Podocnemis expansa	S. American river turtle	16.2	f	[2]
	Podocnemis unifilis	Yellow-spotted river turtle	8.8	m	[2]
Platysternidae	Platysternon megacephalum	Big-headed turtle	15.0	m	[2]
Scincidae	Acontias meleagris	S. African spotted slow skink	3.9	x	[5]
	Chalcides ocellatus	Ocellated skink	3.7	x	[2]
	Chalcides ocellatus	Palaearctic eyed skink	8.9	x	[5]
	Corucia zebrata	Solomon Island skink	2.8	f	[2]
	Egernia bungana		8.6	x	[2]
	Egernia cunninghami	Cunningham's skink	10.4	x	[2]

	Egernia cunninghami	Cunningham's Australian skink	19.9	x	[5]
	Egernia cunninghami	Cunningham's skink	20.0	X	[7]
	Egernia hosmeri	Hosmer's skink	3.3	x	[2]
	Egernia major	Land mullet	10.8	x	[2]
	Egernia stokesii	Stoke's Australian skink	3.0	x	[5]
	Egernia striolatus	Australian striolated skink	3.3	x	[5]
	Egernia whitei	White's Australian skink	3.2	x	[5]
	Eumeces algeriensis		9.0	x	[2]
	Eumeces algeriensis	Algerian skink	3.8	x	[5]
	Eumeces laticeps	Broadhead skink	7.7	f	[2]
	Eumeces obsoletus	Great Plains skink	6.2	x	[2]
	Eumeces schneideri	Schneider's skink	9.3	m	[2]
	Eumeces schneideri	Gold skink	7.5	x	[5]
	Lygosoma cauarinae		5.3	x	[2]
	Lygosoma decresiense	Australian descresian skink	2.4	x	[5]
	Lygosoma labillardierii	Labillarier's Australian skink	2.0	x	[5]
	Lygosoma lesueurii	Lesueur's Australian skink	8.9	x	[5]
	Lygosoma lineo-ocellatum	New Zealand skink	2.3	x	[5]
	Lygosoma quoyi	Quoy's Australian skink	3.0	x	[5]
	Macroscincus cocteauii	Cocteau's Cape Verde Is. skink	8.9	x	[5]
	Ophiomorus tridactylus	Three-toad snake skink	5.8	x	[2]
	Riopa fernandi		1.5	x	[2]
	Tiliqua gigas	Great moluccan skink	9.8	x	[5]
	Tiliqua nigrolutea	Australian black and yellow skink	11.5	x	[5]
	Tiliqua scincoides	Eastern blue tongue skink	9.0	f	[2]
	Tiliqua scincoides	Australian blue-tongued lizard	14.3	x	[5]
	Trachydosaurus rugosus		14.5	x	[2]
	Trachysaurus rugosus	Australian stump-tailed skink	7.1	x	[5]
Sphenodontidae	Sphenodon punctatus	Tuatara	7.0	m	[2]
	Sphenodon punctatus	New Zealand tuatara	11.7	x	[5]
	Sphenodon punctatus	Tuatara	35.0	x	[4]
	Sphenodon punctatus	Tuatara	50.0	x	[3]
	Sphcnodon punctatus	Tuatara	77.0	x	[7]
Teiidae	Ameiva ameiva	Giant ameiva	2.8	m	[2]
	Cnemidophorus neomexicanus	New Mexico whiptail	3.2	f	[2]
	Cnemidophorus tigris	Western whiptail	7.8	x	[10]
	Dracaena guianensis	Guyana caiman lizard	9.3	f	[2]
	Tupinambis nigropunctatus		11.9	m	[2]
	Tupinambis nigropunctatus	S. American black-pointed tegu lizard	5.3	x	[5]
	Tupinambis rufescens	Red tegu	10.6	m	[2]
	Tupinambis rufescens	S. American red tegu lizard	7.3	x	[5]
	Tupinambis teguexin	S. American tegu lizard	12.9	x	[5]
	Tupinambis teguixin	Black tegu	13.0	x	[2]
	Tupinambis teguixin	Black tegu	7.6	x	[2]
Testudinidae	Batagur baska	Indian batagur terrapin	9.2	x	[5]
	Batagur baska	River terrapin	6.1	f	[2]
	Chrysemys concinna suwanniensis		40.7	m	[2]
	Chrysemys floridana		12.5	x	[2]
	Chrysemys nelsoni		15.8	x	[2]
	Chrysemys ornata	Central American adorned terrapin	14.3	x	[5]

Chrysemys picta	Painted turtle	30.0	x	[9] [10]
Chrysemys picta dorsalis	Southern painted turtle	16.6	x	[2]
Chrysemys picta marginata	Midland painted turtle	16.2	f	[2]
Chrysemys picta picta	Eastern painted turtle	14.0	f	[2]
Chrysemys rubriventris		4.3	f	[2]
Chrysemys scripta elegans		30.5	x	[2]
Chrysemys scripta	N. American serrated terrapin	6.8	x	[5]
Chrysemys scripta elegans	N. American elegant terrapin	5.9	x	[5]
Chrysemys scripta rugosa	N. American wrinkled terrapin	12.4	x	[5]
Chrysemys scripta hiltoni		25.3	x	[2]
Chrysemys scripta scripta		8.6	f	[2]
Chrysemys stejnegeri granti		8.1	f	[2]
Cinixys belliana	African tortoise	10.8	x	[5]
Cistudo carolina	N. American box tortoise	35.0	x	[5]
Cuora amboinensis	Southeast Asian box turtle	38.2	x	[2]
Cuora trifasciatta	Chinese three-striped box turtle	9.6	m	[2]
Cyclemus amboinensis	Amboina box tortoise	19.1	x	[5]
Cyclemus dhor	Oldham's East Indian terrapin	9.1	x	[5]
Cyclemus platynota	East Indian flat-backed terrapin	11.8	x	[5]
Cyclemus trifasciata	Chinese trifasciated terrapin	10.8	x	[5]
Geochelone carbonaria	Redfoot tortoise	13.8	f	[2]
Geochelone chilensis	Chaco tortoise	19.8	f	[2]
Geochelone denticulata	S. American yellowfoot tortoise	14.1	m	[2]
Geochelone elegans	Indian star tortoise	14.1	f	[2]
Geochelone elephantopus		47.6	m	[2]
Geochelone gigantea	Aldabra tortoise	37.3	x	[2]
Geochelone gigantea	Aldabra tortoise	152.0	x	[3] [7]
Geochelone gigantea	Aldabra tortoise	70.0	x	[1] [10]
Geochelone pardalis	Leopard tortoise	7.3	f	[2]
Geochelone radiata	Radiated tortoise	12.8	m/f	[2]
Geochelone travancorica		11.8	f	[2]
Geoclemys reevesii	Reeve's Chinese terrapin	23.5	x	[5]
Geoemyda grandis		20.4	x	[2]
Geoemyda pulcherrima pulcherrima		20.4	m	[2]
Geoemyda rubida		11.8	x	[2]
Geoemyda spinosa	East Indian spinous terrapin	3.2	x	[5]
Geoemyda spinosa		9.3	x	[2]
Geoemyda trijuga		40.0	x	[2]
Gopherus agassizi	Desert tortoise	2.8	f	[2]
Gopherus berlandieri	Texas tortoise	4.4	m	[2]
Gopherus flavomarginata	Bolson tortoise	4.8	f	[2]
Gopherus polyphemus	Gopher tortoise	8.5	x	[2]
Homopus areolatus	Beaked cape tortoise	7.3	f	[2]
Homopus areolatus	African tortoise	3.1	x	[5]
Kachuga dhongoka	Indian dhongoka terrapin	6.8	x	[5]
Kachuga smithii	Brown roofed turtle	9.8	m	[2]
Kachuga tecta tentoria	Indian roof turtle	11.0	f	[2]
Kachuga tecta	Indian roofed terrapin	3.3	x	[5]
Kinixys belliana belliana	Bell's hingeback tortoise	14.4	x	[2]
Kinixys erosa	Serrated hingeback tortoise	10.3	m	[2]

	Malacochersus tornieri	African pancake tortoise	7.3	x	[2]
	Nicoria punctularia	S. American rough terrapin	5.8	x	[5]
	Nicoria trijuga	Ceylonese terrapin	12.8	x	[5]
	Testudo angulata	S. African angulated tortoise	11.8	x	[5]
	Testudo elongata	Burmese tortoise	11.8	x	[5]
	Testudo emys	East Indian upland tortoise	11.9	x	[5]
	Testudo graeca	Mediteranean spur-thighed tortoise	4.7	x	[2]
	Testudo hermanni	Heermann's tortoise	8.3	f	[2]
	Testudo microphyes	Galapagos albemarle tortoise	16.5	x	[5]
	Testudo pardalis	African leopard tortoise	10.1	x	[5]
	Testudo radiata	Madagascar radiated tortoise	9.8	x	[5]
	Testudo tabulata	Brazilian tortoise	16.2	x	[5]
Trionychidae	Cyclanorbis senegalensis	Senegal soft turtle	6.4	x	[5]
	Lissemys punctata	Indian flapshell turtle	12.8	x	[2]
	Pelochelys bibroni	Asian giant softshell	10.1	x	[2]
	Trionyx cartilagineus		7.3	m	[2]
	Trionyx gangeticus	Ganges soft turtle	14.0	x	[5]
	Trionyx spinifer	N. American spinous soft turtle	10.6	x	[5]
	Trionyx spiniferus spiniferus	Spiny softshell	16.8	x	[2]
	Trionyx triunguis	African softshell	25.2	m	[2]
	Trionyx triunguis	African softshell	25.0	x	[7]
Typhlopidae	Typhlops delalandii	Delalande's ground snake	0.8	x	[5]
	Typhlops punctatus	Punctuated ground snake	0.3	x	[5]
Varanidae	Lanthanotus borneensis	Borneo earless monitor	6.9	x	[2]
	Varanus acanthurus brachyurus	Ridgetail monitor	10.0	f	[2]
	Varanus bengalensis	Bengal monitor	10.0	x	[2]
	Varanus flavescens	Yellow monitor	4.5	m	[2]
	Varanus flavescens	Indian yellowish waral lizard	4.8	x	[5]
	Varanus giganteus	Perentie	5.9	x	[2]
	Varanus gouldi	Sand monitor	7.8	x	[2]
	Varanus gouldi	Gould's Australian waral lizard	6.8	x	[5]
	Varanus griseus	Desert monitor	4.5	m	[2]
	Varanus griseus	Desert waral lizard	9.3	x	[5]
	Varanus komodoensis	Komodo dragon	8.9	f	[2]
	Varanus mertensi	Merten's water monitor	6.6	x	[2]
	Varanus mitchelli	Mitchell's water monitor	6.6	x	[2]
	Varanus niloticus	Nile monitor	6.6	x	[2]
	Varanus niloticus	Nile waral lizard	6.4	x	[5]
	Varanus niloticus	Nile waral lizard	5.5	m	[5]
	Varanus salavator	Crocodile monitor	12.3	m	[2]
	Varanus salvator	East Indian great water lizard	10.6	x	[5]
	Varanus spenceri	Spencer's monitor	4.3	x	[2]
	Varanus timorensis	Spotted tree monitor	6.8	x	[2]
	Varanus tristis	Arid monitor	8.3	x	[2]
	Varanus varius	Lace monitor	9.5	m	[2]
	Varanus varius	Australian lace lizard	6.8	x	[5]
	Varanus varius	Lace monitor	7.0	x	[7]
Viperidae	Agkistrodon acutus		7.5	m	[2]
	Agkistrodon bilineatus bilineatus	Cantil	24.3	x	[2]
	Agkistrodon contortrix	Northern copperhead	14.5	m	[2]

contortrix				
Agkistrodon contortrix laticinctus	Broad-banded copperhead	21.5	m	[2]
Agkistrodon contortrix mokasen	Northern copperhead	29.8	f	[2]
Agkistrodon controtrix pictigaster	Trans-pecos copperhead	20.4	m	[2]
Agkistrodon halys blomhoffi	Siberian pit viper	6.2	f	[2]
Agkistrodon halys brevicaudus	Siberian pit viper	12.5	m	[2]
Agkistrodon hypnale		7.3	x	[2]
Agkistrodon piscivorus	Water moccasin	21.0	x	[7]
Agkistrodon piscivorus conanti	Florida cotton-mouth	15.7	x	[2]
Agkistrodon piscivorus leucostoma	Western cotton-mouth	16.5	m	[2]
Agkistrodon piscivorus piscivorus	Eastern cotton-mouth	13.1	m	[2]
Agkistrodon piscivorus ssp.	Cotton-mouth	18.9	x	[2]
Agkistrodon rhodostoma		5.8	f	[2]
Atheris squamiger	Rough-scaled bush viper	4.5	x	[2]
Atractaspis bibronii	Bibron's mole viper	8.9	x	[2]
Bitis arietans arietans	Puff adder	15.8	m	[2]
Bitis caudalis	Horned puff adder	5.3	x	[2]
Bitis gabonica gabonica	Gaboon viper	10.8	m	[2]
Bitis gabonica rhinoceros	Gaboon viper	13.3	f	[2]
Bitis nasicornis	Rhinoceros viper	8.3	f	[2]
Bothrops atrox asper	Common lancehead	20.2	x	[2]
Bothrops atrox atrox	Common lancehead	8.5	x	[2]
Bothrops godmani		7.4	x	[2]
Bothrops jararaca	Jararaca	6.5	x	[2]
Causus rhombeatus	Cape viper	6.6	x	[5]
Cerastes cerastes	Horned viper	17.0	f	[2]
Cerastes vipera	Common sand viper	6.0	x	[2]
Crotalus adamanteus	Eastern diamondback rattlesnake	22.8	x	[2]
Crotalus atrox	Western diamondback rattlesnake	25.8	f	[2]
Crotalus atrox	Western diamondback rattlesnake	22.0	x	[7]
Crotalus basiliscus basiliscus	Mexican West-Coast rattlesnake	12.8	f	[2]
Crotalus catalinensis	Santa Catalina Island rattlesnake	10.3	f	[2]
Crotalus cerastes cerastes	Mojave Desert sidewinder	8.1	m	[2]
Crotalus cerastes cercobombus	Sonoran sidewinder	10.7	m	[2]
Crotalus cerastes laterorepens	Colorado Desert sidewinder	13.8	x	[2]
Crotalus durissus durissus	Neotropical rattlesnake	12.0	x	[2]
Crotalus durissus terrificus	Neotropical rattlesnake	13.4	x	[2]
Crotalus enyo enyo	Baja California rattlesnake	17.1	x	[2]
Crotalus horridus	Timber rattlesnake	30.2	m	[2]
Crotalus lepidus klauberi	New Mexico banded rock rattlesnake	23.3	x	[2]
Crotalus lepidus lepidus	Mottled rock rattlesnake	17.9	x	[2]
Crotalus lepidus maculosus	Rock rattlesnake	7.2	m	[2]
Crotalus mitchelli pyrrhus	Southwestern speckled rattlesnake	15.5	f	[2]
Crotalus mitchelli stephensi	Panamint rattlesnake	12.4	f	[2]
Crotalus molossus molossus	Northern black-tailed rattlesnake	15.5	m	[2]
Crotalus molossus nigrescens		11.2	x	[2]

Crotalus polystictus	Mexican lancehead rattlesnake	5.7	m	[2]
Crotalus pricei miquihuanus	Twin-spotted rattlesnake	5.1	m	[2]
Crotalus pricei pricei	Twin-spotted rattlesnake	3.5	m	[2]
Crotalus pusillus	Tancitanan dusky rattlesnake	5.7	x	[2]
Crotalus ruber ruber	Northern red rattlesnake	14.4	m	[2]
Crotalus scutulatus scutulatus	Mojave rattlesnake	13.0	f	[2]
Crotalus tigris	Tiger rattlesnake	15.3	m	[2]
Crotalus tortugensis	Tortuga Island rattlesnake	18.3	m	[2]
Crotalus triseriatus triseriatus	Mexican dusky rattlesnake	7.0	f	[2]
Crotalus unicolor	Arulon Island rattlesnake	14.8	x	[2]
Crotalus viridis cerberus	Arizona black rattlesnake	12.4	x	[2]
Crotalus viridis helleri	Southern Pacific rattlesnake	19.4	m	[2]
Crotalus viridis lutosus	Great Basin rattlesnake	17.1	m	[2]
Crotalus viridis nuntius	Hopi rattlesnake	6.2	m	[2]
Crotalus viridis oreganus	Northern Pacific rattlesnake	11.0	x	[2]
Crotalus viridis viridis	Prairie rattlesnake	19.3	m	[2]
Crotalus willardi willardi	Arizona ridge-nosed rattlesnake	21.3	f	[2]
Echis carinatus	Saw-scaled viper	11.8	f	[2]
Echis coloratus	Palestine saw-scaled viper	11.8	f	[2]
Eristocophis macmahonii	McMahon's viper	9.3	m	[2]
Lachesis muta stenophrys	Bushmaster	8.5	x	[2]
Pelamis platurus	Yellow-bellied sea snake	2.1	f	[2]
Sistrurus catenatus catenatus	Massasauga	9.9	f	[2]
Sistrurus catenatus ssp	Massasauga	14.0	x	[2]
Sistrurus catenatus tergeminus	Western massasauga	12.8	x	[2]
Sistrurus miliarius barbouri	Dusky pygmy rattlesnake	15.1	m	[2]
Sistrurus miliarius miliarius	Carolina pygmy rattlesnake	7.2	x	[2]
Sistrurus miliarius streckeri	Western pygmy rattlesnake	6.6	x	[2]
Sistrurus ravus	Mexican pygmy rattlesnake	10.0	m	[2]
Trimeresurus albolabris	White-lipped tree viper	9.3	f	[2]
Trimeresurus elegans	Elegant pit viper	10.3	x	[2]
Trimeresurus flavoviridis flavoviridis	Yellow-spotted pit viper	8.9	x	[2]
Trimeresurus gramineus	Indian green tree viper	6.2	x	[2]
Trimeresurus monticola		3.9	m	[2]
Trimeresurus okinavensis		10.8	x	[2]
Trimeresurus purpureomaculatus	Mangrove viper	8.9	x	[2]
Trimeresurus steonegeri	Chinese green tree viper	7.0	x	[2]
Trimeresurus trigonocephus	Ceylon pit viper	5.2	x	[2]
Trimeresurus wagleri		5.8	f	[2]
Ttimeresurus popeorum	Pope's tree viper	6.7	x	[2]
Vipera ammodytes	Long-nosed viper	22.0	x	[7]
Vipera ammodytes ammodytes	Long-nosed viper	9.3	m	[2]
Vipera ammodytes montandoni	Sand viper	8.8	x	[2]
Vipera ammodytes ssp.	Sand viper	7.1	x	[2]
Vipera aspis	Asp viper	7.8	m	[2]
Vipera aspis	European asp-viper	3.9	x	[5]
Vipera berus	Northern palaearctic viper	2.1	x	[5]
Vipera latasti	Iataste's viper	5.5	f	[2]
Vipera lebetina obtusa	Bluntnose viper	10.1	m	[2]

	Vipera lebetina schwezeri	Bluntnose viper	11.3	x	[2]
	Vipera lebetina ssp.	Bluntnose viper	11.3	x	[2]
	Vipera lebetina turanica	Bluntnose viper	5.1	x	[2]
	Vipera russelli russellii	Russell's viper	13.3	x	[2]
	Vipera russelli ssp.	Russell's viper	11.6	m	[2]
	Vipera santhina raddei		6.1	m	[2]
	Vipera xanthina palestinae	Coastal viper	10.1	m	[2]
	Walterinnesia aegyptia	Black desert cobra	6.7	m	[2]
Xantusiidae	Klauberina riversiana	Island night lizard	5.3	m	[2]
	Lepidophyma flavimaculatum flavimaculata	Yellow-spotted night lizard	9.9	f	[2]
	Xantusia vigilis	Desert night lizard	3.8	x	[2]
Xenopeltidae	Xenopeltis unicolor		9.3	x	[2]
Zonuridae	Pseudocordylus microlepidotus	S. African smooth-backed lizard	2.8	x	[5]
	Zonurus cordylus	S. African rough-scaled girdled lizard	3.5	x	[5]
	Zonurus giganteus	S. African derbian girdled lizard	4.5	x	[5]

References

1. Bourne, D. and M. Coe, 1978. The size, structure and distribution of the giant tortoise population of Aldabra. *Philosophical Transactions Royal Society London, Series B* 282: 139-175.
2. Bowler, J.K., 1975. Longevity of reptiles and amphibians in N. American collections as of 1 November, 1975. *Society for the Study of Amphibians and Reptiles, Miscellaneous Publications, Herpetological Circular* 6:1-32.
3. Burton, M. and R. Burton, 1975. *Encyclopedia of Reptiles, Amphibians and Other Cold-Blooded Animals*. New York: BPC Publishing Ltd.
4. Castanet, J., D.G. Newman, and H. Saint-Girons, 1988. Skeletochronological data on the growth, age, and population structure of the tuatara, *Sphenodon punctatus*, on Stephens and Lady Alice Islands, New Zealand. *Herpetologica* 44: 25-37.
5. Flower, M.S.S., 1925. Contributions to our knowledge of the duration of life in vertebrate animals - III. Reptiles. *Proceedings of the Zoological Society of London* 1925(2): 911-981.
6. Frazer, N.B., J.W. Gibbons, and J.L. Greene, 1990. Life tables of a slider turtle population, in *Life History and Ecology of the Slider Turtle*, J.W. Gibbons, Editor. Washington, D.C.: Smithsonian Institute Press. .
7. Goin, C.J., O.B. Goin, and G.R. Zug, 1978. *Introduction to Herpetology,* 3rd ed. San Francisco: W. H. Freeman and Company.
8. Tyning, T.F., 1990. *A Guide to Amphibians and Reptiles,* 1 ed. Boston: Little, Brown and Company.
9. Wilbur, H.M., 1975. The evolutionary and mathematical demography of the turtle, *Chrysemys picta*. *Ecology* 56: 64-77.
10. Zug, G.R., 1993. *Herpetology: An Introductory Biology of Amphibians and Reptiles*. San Diego: Academic Press.

Fishes

Fishes are the earliest and largest group of all the back-boned animals and are found throughout the oceans from the tropics through the polar regions and from surface habitats to the deepest oceanic trenches 10 or more kilometers deep. They range in size from 1 centimeter gobies to 12 meter whale sharks (Paxton and Eschmeyer 1995). Fishes are the most diverse vertebrates with over 24,000 species in 4,200 genera, 483 families and 56 orders. The jawless fishes such as the hagfish and lampreys and the cartilaginous fishes, including the sharks and rays, are the most primitive groups and the bony fishes are more advanced. Of this large group of bony fishes, the fleshy-finned fishes (Sarcopterygii) such as the lungfishes and coelacanths are among the oldest and the ray-finned fishes (Actinopterygii) ranging from sturgeons and paddlefish to salmon and perch are among the most recent.

CLASS AGNATHA (Jawless fishes—lampreys and allies)

Lampreys and hagfishes (Agnatha) are not true fishes because they lack paired fins and jaws. However they are usually included with the fishes because they are frequently collected with fishes (Moyle 1993). Lampreys (Order Petromyzontiformes) are long and eel-like as adults that are readily recognizable by the sucking disc on the mouth that is covered with teeth, the line of gill slits along each side of the head and the large eyes. The lamprey latches on to the side of its victim, holding on with its sucking disc and rasping a hole with its tongue. It then sucks out blood and other bodily fluids. Lampreys are found in the temperate waters of the northern and southern hemispheres, and in cool deep waters in some parts of the tropics.

CLASS CHONDRICHTHYES (Cartilaginous fishes—sharks, rays, skates, chimaeras)

Fishes in the Class Chondrichthyes have a skeleton that is almost entirely cartilage and thus they are typically referred to as the cartilaginous fishes. They differ from the Agnatha in that they have well-developed jaws, paired nostrils and paired pectoral and pelvic fins. Longevity records include species in the following orders (descriptions of orders taken from Paxton and Eschmeyer 1995; Moyle 1993): (1) Carcharhiniformes (ground sharks)—this order contains 8 families and over 200 species. They include "typical" sharks such as whaler or requiem sharks, cat sharks and hammerhead sharks; (2) Heterodontiformes (horn sharks)—this order contains one family of 8 species; most species live mainly in shallow-water and are bottom living with restricted distributions in warm temperate and tropical areas; (3) Lamniformes (basking sharks)—(4) Orectoblobiformes (carpet sharks)—a diverse group of 7 families and around 33 species; includes the gigantic whale shark, the carpet shark and the zebra shark. Habitats range from specialist feeders on reefs to open water filter feeders; (5) Rajiformes (skates)—the skates are a diverse group of rays with approximately 200 species worldwide. They live near the bottom in all oceans from shallow estuarine to deepsea habitats; (6) Squaliformes (dogfish sharks)—this is a large group of mainly deepwater sharks which occur in tropical, temperature and even artic waters. There are 3 families and over 90 species; (7) Myliobatiformes (develrays and allies)—this order consists of three families and around 150 species, most of which live on the continental shelf in tropical regions. Species include the plankton-feeding manta which can grow up to 6.5 meters in width and the stingray.

CLASS OSTEICHTHYES (Bony fishes)

This group contains the vast majority of extant species with over 20,000. Longevity records were obtained for species in the following orders (descriptions of orders taken from Paxton and Eschmeyer 1995; Moyle 1993): (1) Acipenseriformes (sturgeons and paddlefish)—confined to the northern hemisphere in both fresh waters and coastal marine waters; sturgeons are bottom feeders and paddlefish are filter-feeders; (2) Amiformes (mudfish; bowfins)—typical lowland species common in backwaters and oxbow lakes preferring clear waters; (3) Anguilliformes (eels)—this group consists of 15 families with about 650 species; includes the tropical and subtropical moray and conger eels and the freshwater eels that live in the Americas and Europe; (4) Atheriniformes (silverside family)—this group consists of 7 families with approximately 275 species; includes California grunion and jacksmelt; (5) Ceratodonitiformes (lungfish)—species in this group are relicts of ancient fish groups that were near the evolutionary stem of amphibians. They possess a swim bladder that developed into a highly vascularized lung that enables them to breathe air; (6) Characiformes (Charcoid)—this group consists of 12 families and around 1,400 species found in Africa and Central and South America; best known species is the South American piranha; (7) Clupeiformes (shad; herring)—(8) Cyprinidontiformes (killifishes and ricefishes)—this group consists of 9 families and 7000-800 species; members in this group are among the most popular fishes in hobby aquariums including the guppy, the platy and *Gambusia* spp; (9) Cypriniformes (carps; loaches; shiners)—this mostly freshwater group consists of 6 families and over 2,000 species; found on all continents capable of supporting fish life; (10) Gadiformes (cod; haddock)—wide-ranging group found around the world inhabiting deep to shallow seas and ranging from tropics to both polar regions; includes rattails, hakes, and a variety of cod species; (11) Syngnathiformes (seahorses)—seahorses are contained in a single genus, *Hippocampus*; possess a prehensile tail and a head positioned more or less at a right angle to the body—hence the term "seahorses"; (12) Gobiesociformes (dragonettes); (13) Perciformes (perches and allies)—found in almost all kinds of aquiatic habitats from mountain streams to deep oceans; over 9,000 species in 160 different families including drums, croakers, snappers, butterfly fish, freshwater perch, sunfishes, freshwater basses, tunas, mackerals and billfishes; (14) Pleuronectiformes (flatfishes; soles; flounders)—this group consists of 7 families and 540 species; found mainly on the continental slopes of oceans and include many of the seafood varieties including sole, halibut, turbot, and brill; (15) Salmoniformes (pikes; smelts; salmons and allies)—most salmoniform families occur in fresh water though some species spend part of their cycle in the sea and migrate to rivers to spawn; smelts resemble salmons and trout but are smaller and primarily marine; pikes are mostly freshwater, large, toothed predators; (16) Siluriformes (catfishes)—more than 2,200 species of catfishes are known with extant members found on all continents except Antarctica; found in almost all freshwater environments ranging from mountain streams to slow-moving rivers; some species found in coastal marine habitats.

References Cited

Moyle, P. B., 1993. *Fish: An Enthusiast's Guide*. Berkeley, University of California Press.

Paxton, J. R. and W. N. Eschmeyer, 1995. *Encyclopedia of Fishes*. San Diego, Academic Press.

Table 4. Record Life Spans (years) of Fishes

((+) indicates that individual was still alive at the time that life span was recorded)

Order/Family	Genus/Species	Common Name	Wild	Capt.	M/F	Reference
SUBPYLUM CEPHALOCHORDATA						
CLASS BRACHIOSTOMATA						
Amphioxiformes						
Branchiostomidae	Branchiostoma lanceolatum	Lancelet/ amphioxus		0.6	x	[70]
SUBPYLUM CHORDATA						
CLASS CEPHALOSPIDOMORPHI						
Petromyzoniformes						
Petromyzonidae	Ichthyomyzon fossor		6.0		x	[36] [3]
	Lampetra fluviatilis	Lamprey (+)		1.0	x	[70] [3]
	Lampetra fluviatilis	River-lamprey or lampern		0.4	x	[70]
	Lampetra tridentata	Pacific lamprey	9.0		x	[177] [143] [198]
	Lampetra wilderli	N. American brook-lamprey		0.5	x	[70]
	Petromyzon branchialis	Fringe-lipped lampern		0.2	x	[70]
	Petromyzon marinus	Sea-lamprey		0.8	x	[70]
	Petromyzon marinus	Sea-lamprey	7.0		x	[200] [3]
CLASS CHONDRICHTHYES						
Carchariniformes						
Carchardinae	Carcharhinus leucas	Bull shark	13.7		x	[81]
	Carcharhinus leucas	Bull shark	28.0		x	[99]
	Carcharhinus obscurus	Dusky shark	30.0		x	[99]
	Galeocerdo cuvier	Tiger shark		3.3	x	[53]
	Triaenodan obesus	Whitetip reef shark	25.0		x	[160]
Scyliorhinidae	Scyliorhinus caniculus	Cat shark	10.0		x	[48]
	Scyliorhinus stellaris	Cat shark	19.0		x	[48]
	Scyliorhinus stellaris	Cat shark		18.0	x	[71] [3]
Triakidae	Triakis semifasciata	Leopard shark (+)	14.0		x	[1] [51] [59]
	Triakis semifasciata	Hound shark	10.0		x	[48]
	Galeorhinus australis	Hound shark	27.6		x	[81]
	Galeorhinus japonicus	Eiraku shark	15.0		x	[193]
Heterodontiformes						
Heterodontidae	Heterodontis francisci	Horned shark	12.0		x	[48] [98]
	Heterodontis galeatus	Shark	12.0		x	[48]
Lamniformes						
Cetorhinidae	Cetorhinus maxims	Basking shark	32.0		x	[193]
Orectolobiformes						
Ginglymostomatidae	Ginglymostoma cirratum	Nurse shark	25.0		x	[48]
Hemiscyllidae	Chiloscyllium plagiosum	Bamboo shark	25.0		x	[48]
Orectolobidae	Ginglymostoma spp.	Carpet shark		9.0	x	[71] [3]
Rhiniodontidae	Rhiniodon typus	Whale shark	70.0		x	[30]
Rajiformes						
Dasyatidae	Dasyatis pastinaca	Sting rays (+)		21.0	x	[71] [3]
Rajidae	Raja clavata	Skate	12.0		x	[111] [52]

	Raja maculata	Skate (+)		5.8	x	[77] [3]
Squaliformes						
Squalidae	Squalus acanthias	Spur dogfish	60.0		x	[28]
CLASS OSTEICHTHYES						
Acipenseriformes						
Acipenseridae	Acipenser brevirostris		14		x	[3]
	Acipenser brevirostrum	Short-nosed sturgeon		7	x	[70]
	Acipenser fulvescens	Lake sturgeon	152		x	[5]
	Acipenser fulvescens	Lake sturgeon	82		x	[30] [156]
	Acipenser medirostris	Green sturgeon	60		x	[97]
	Acipenser medirostris	Green sturgeon	80		x	[158]
	Acipenser oxyrhynchus		12		x	[3]
	Acipenser ruthenus	Sturgeon		46.1	x	[71]
	Acipenser sturio	Sturgeon (+)		6.8	x	[70][71] [3]
	Acipenser transmontanus	White sturgeon	82		x	[3]
	Acipenser transmontanus	White sturgeon (+)	100		x	[7]
	Acipenser transmontanus	White sturgeon	80		x	[158]
	Huso huso	Beluga sturgeon	118		x	
Polydontidae	Polyodon spathula	American paddlefish	24		x	[179] [3]
	Polyodon spathula	American paddlefish	30		x	[157]
Amiformes						
Amiidae	Amia calva	American mudfish	30		x	[16] [3]
	Amia calva	Bowfin		24	x	[71] [3]
	Amia calva	American mudfish		30	x	[32]
	Amia calva	American mudfish or bowfin		20	x	[70]
Anguilliformes						
Anguilidae	Anguilla anguilla	European eel	88		x	[30]
	Anguilla anguilla	Freshwater eel		55	x	[71] [3]
	Anguilla anguilla	Eel	17		x	[77]
	Anguilla chrisypa	N. American eel		6	x	[70]
	Anguilla rostrata	Eel		50	x	[70]
	Anguilla vulgaris		15		x	[3]
Congridae	Conger conger	Conger eel		9	x	[70] [3]
	Conger myriaster	Brevoort	8		x	[192] [3]
	Muraenesox cinereus		15		x	[3]
Atheriniformes						
Atherinidae	Atherinops affinis	Topsmelt	8		x	[174] [61]
	Atherinops affinis	Topsmelt	9		x	[174] [62]
	Atherinops californiensis	Jacksmelt	11		x	[75]
	Atherinops californiensis	Jacksmelt	10		x	[44]
	Atherinops californiensis	Jacksmelt	11		x	[19] [3]
	Labidesthes sicculus	Brook silverside	1.3		x	[101]
	Leuresthes tenuis	California grunion (+)	3		f	[43]
	Leuresthes tenuis	California grunion	8		m	[43]
	Leuresthes tenuis	California grunion	8		f	[43]
	Leuresthes tenuis	California grunion	4		x	[45] [3]
	Menidia beryllina	Inland silverside	1		m	[128] [143]

	Menidia beryllina	Inland silverside	2		f	[128] [143]
Ceratodontiformes						
Ceratodontidae	Ceratodus forsteri	Australian lungfish (+)		19.7	x	[70][71] [3]
Characiformes						
Characidae	Alestes spp.	Charcoid (+)		8.3	x	[70] [3]
	Astyanax bimaculatus	Characin	18		x	[98]
	Hydrocyon spp.	Tetra (+)		7.9	x	[70] [3]
	Hydrocyon brevis	African tetra		4.2	x	[70]
	Hydrocyon forskalii	Nile tiger-fish		7.9	x	[70]
Characinidae	Distichodus niloticus	The Nile "lafash"		7.5	x	[70]
Clupeiformes						
Clupeidae	Alosa pseudoharengus		9		x	[3]
	Alosa sapidissima	American shad	13		x	[71] [3]
	Alosa sapidissima	American shad	7		x	[50]
	Alosa sapidissima	American shad (+)	5		x	[143]
	Clupea harengus	Atlantic herring	19		x	[71]
	Clupea harengus	Atlantic herring	22		x	[115] [168]
	Clupea harengus pallasi	Pacific herring	9		x	[132]
	Clupea pallasi	Pacific herring	19		x	[3] [91]
	Clupea pallasi	Pacific herring	11		x	[194]
	Dorosoma petenense	Threadfin shad	4		x	[103]
	Sardinops caerulea	California sardine	13		x	[183] [3]
	Sardinops caerulea	California sardine	18		x	[46] [182] [152] [28]
	Sardinops neopilchardus	Australian sardine	6.5		x	[29]
Dorosomidae	Dorosoma cepedianum		10		x	[3]
Engraulidae	Anchoa compressa	Deepbody anchovy	6		x	[94]
	Anchoa delicatissima	Slough anchovy	3		x	[94]
	Engraulis mordax	Northern anchovy	7		x	[47] [3] [21] [75]
Coelacanthiformes						
Protopteridae	Protopterus annectens	American lungfish		18	x	[70, 71] [3]
	Protopterus annectens	African lungfish		9.8	x	[70]
Cyprinidontiformes						
Cyprinidontidae	Cyprinodon tularosa	White sands pupfish	5		x	[102]
Poeciliidae	Gambusia affinis	Mosquito fish	1.5		x	[112] [3]
Cypriniformes						
Balitoridae	Noemacheilus barbatulus	Stone loach	6		x	[136]
Catostomidae	Carpiodes carpio		6		x	[3]
	Catostomus catostomus		13		x	[162] [3]
	Catostomus commersonii		12		x	[162] [3]
	Catostomus occidentalis	Sacramento sucker	10		x	[143]
	Erimyzon sucetta		8		x	[3]
	Ictiobus bubalus		12		x	[3]
	Ictiobus cyprinellus		14		x	[3]
	Moxostoma aureolum		9		x	[3]
	Thoburnia rhothoeca		7		x	[161] [3]
Cobitidae	Misgurnus anguillicaudatus	Loach (+)		10.2	x	[71] [3]
	Misgurnus fossilis	Loach (+)		21.7	x	[70] [3]
Cyprinidae	Abramis brama	Bream		17	x	[71] [3]
	Abramis brama	Bream	10		x	[178] [3]

	Abramis brama	Bream	23		x	[28]
	Abramis chrysoleucas	Common roach/golden shiner (+)		7	x	[70]
	Abramis chrysoleucas roseus	'Pearl roach' (+)		10	x	[70]
	Barbus bynni	Barb (+)		16.3	x	[70] [3]
	Carassius auratus	Goldfish		30	x	[70] [37] [143]
	Carassius auratus	Goldfish	41		x	[30]
	Couesius plumbeus		4		x	[144] [3]
	Cyprinius carpio	Carp	38		x	[98]
	Cyprinus carassius auratus	Goldfish		10	x	[70]
	Cyprinus carpio	Common carp		47	x	[71]
	Cyprinus carpio	Common carp	20		x	[125]
	Cyprinus carpio	Carp (+)		6	x	[70]
	Cyprinus carpio	Common carp	38		x	[98]
	Cyprinus carpio var.	Prussian carp		6.4	x	[70]
	Hesperoleucus symmetricus	California roach	5		x	[143]
	Hybognathus nuchalis		2		x	[161] [3]
	Hyborhynchus notatus		3		x	[36] [3]
	Idus melanotus		6		x	[178] [3]
	Lavinia exilicauda		6		x	[144] [3]
	Leuciscus cephalus	Chub		11	x	[71] [3]
	Leuciscus idus	Golden ide		29.8	x	[70][71] [3]
	Leuciscus leuciscus	Carp		8	x	[70] [3]
	Leuciscus orfus	Golden orfe (+)		14.25	x	[70]
	Leuciscus rutilus		8		x	[178] [3]
	Notemigonus crysoleucas	Golden shiner	9		x	[52] [3] [37]
	Notemigonus crysoleucas	Golden shiner (+)		7	x	[70] [3]
	Notropis cornutus		6		x	[167] [3]
	Notropis lutrensis	Red shiner	3		x	[37]
	Orthodon microlepidotus		5		x	[145] [3]
	Phoxinus phoxinus	Minnow		13	x	[71] [3]
	Phoxinus phoxinus	Minnow	3		x	[76]
	Pimephales promelas	Fathead minnow (+)	2		x	[177]
	Pogonichthys macrolepidotus	Splittail	5		x	[39]
	Ptychocheilus grandis	Sacramento squawfish	9		x	[191]
	Ptychocheilus oregonensis		11		x	[49] [3]
	Rhinichthys cataractae		5		x	[3]
	Rutilus rutilus		12		x	[71] [3]
	Scardinius erythrophthalmus	Rudd		10	x	[70] [3]
	Semotilus corporalis		6		x	[3]
Electrophoridae	Electrophorus electricus	Electric eel		11.5	x	[70]
Gymnotidae	Gymnotus electricus	Electric eel		12.6	x	[70]
Elopiformes						
Megalopidae	Megalops cyprinoides	Pacific tarpon	44		x	[113]
Esociformes						
Esocidae	Esox lucius	Pike		10	x	[70] [3]
	Esox lucius	Pike	24		x	[133]
	Esox lucius	Pike		6.8	x	[70]
	Esox masquinongy	Chain pickerel	19		x	[36] [3]

	Esox masquinongy	Chain pickerel		10	x	[70] [3]
	Esox niger		8		x	[3]
Gadiformes						
Gadidae	Brosme brosme		14		x	[3]
	Eleginus navaga		12		x	[3]
	Gadus minutus	Poor cod	5		m	[127]
	Gadus minutus	Poor cod	5		f	[127]
	Gadus morhua	Atlantic cod	16		x	[69]
	Gadus virens	Coalfish	23		x	[172] [3]
	Lota lota		16		x	[162] [3]
	Melanogrammus aeglefinus	Haddock	15		x	[71] [3]
	Melanogrammus aeglefinus	Haddock	10		x	[28]
	Merlangius merlangus		10		x	[129] [3]
	Merluccius merluccius	Hake	12		x	[71] [3]
	Molva molva		14		x	[3]
	Phycis blennoides		10		x	[92] [3]
	Theragra chalcogramma		15		x	[3]
Merlucciidae	Urophycis tenuis	White hake	23		x	[28]
Gasterosteiformes						
Gasterosteidae	Aulorhynchus flavidus	Tube-snout	1		x	[68]
	Gasterosteus aculeatus	Three-spine stickleback		2.4	x	[71] [3]
	Gasterosteus aculeatus	Three-spine stickleback	3		x	[83] [3] [143]
	Gasterosteus aculeatus	Three-spine stickleback	8		x	[165]
	Gasterosteus aculeatus	Eight-spine stickleback, Ten-spine stickleback	4		x	[105]
Syngnathidae	Hippocampus antiguorum	Short-nosed sea horse		1.3	x	[70]
	Hippocampus guttulatus	Sea horse (+)		6	x	[70] [3]
	Hippocampus hippocampus	Sea horse (+)		1.3	x	[70] [3]
	Hippocampus hudsonius	Hudson sea horse (+)		4.67	x	[70]
	Hippocampus hudsonius	Hudson sea horse		4.6	x	[70]
	Hippocampus hudsonius	Sea horse, and pygmy sea horse	1		x	[95] [187]
	Hippocampus ramulosus	Branched sea horse		0.5	x	[70]
Gobiesociformes						
Callionymidae	Callionymus lyra	Dragonette	7		x	[40] [3]
	Callionymus lyra	Dragonette	4		m	[40]
	Callionymus lyra	Dragonette	6		f	[40]
Hemirhamphiformes						
Hemirhamphidae	Reporhamphus melanochir	Sea gar fish	7		x	[119] [3]
Lepidosireniiformes						
Lepidosirenidae	Lepidosiren paradoxa	S. American lungfish (+)		8.25	x	[71] [3]
	Lepidosiren paradoxa	S. American lungfish	8.3		x	[71]
Lophiiformes						
Lophiidae	Lophius americanus	Angler	30		x	[28]
Mugliformes						
Mugilidae	Mugil cephalus	Mullet		11	x	[71] [3]
Myctophiformes						
Myctophidae	Lampanyctus leucopsaurus		6.0		x	[3]

	Stenobrachius leucopsarus	Northern lampfish	8		x	[91]
Osteoglossiformes						
Mormyridae	Gnathonemus cyprinoides	Elephantfish		7.1	x	[70]
	Hyperopisus bebe	Elephantfish (+)		10.1	x	[70] [3]
	Marcusenius isidori	Mormyrid (+)		28	x	[71] [3]
	Marcusenius isidori	Mormyrid	29		x	[98]
	Mormyrus cashive	Elephantfish		6.3	x	[76]
	Mormyrus kannume	Elephant-snout fish (+)		6.3	x	[70]
Notopteridae	Xenomystus nigri	Knifefish		11.4	x	[70] [3]
Osteoglossidae	Osteoglossam bicirrhosum	Arowana	6.5		x	[98]
Perciformes						
Ammodytidae	Ammodytes hexapterus	Pacific sand lance	8		x	[67]
Anabantidae	Anabas kingsleyeae	Miss Kingsley's climbing perch		8.7	x	[70]
	Anabas scandens	Climbing perch (+)		8	x	[70]
	Anabus kingsleyeae	Gourami		8.7	x	[70] [3]
	Anabus testudineus	Gourami		11	x	[71] [3]
	Macropodus opercularis	Gourami (+)		8	x	[70] [3]
Blenniidae	Blennius gattorugin	Combtooth blenny		5	x	[70][71] [3]
	Blennius tolis	Blenny	6		x	[159]
Carangidae	Seriola dorsalis	Yellowtail sardine	12		x	[20] [3]
	Seriola dorsalis	Jack		7.25	x	[154] [3]
	Trachurus symmetricus	Peruvian horse mackerel	30		x	[3]
Centrarchidae	Ambloplites rupestris	Rock bass	18		x	[3]
	Ambloplites rupestris	Perch		12	x	[70][71] [3]
	Ambloplites rupestris	N. American rock-bass (+)		7	x	[70]
	Archoplites interruptus	Sacramento perch	9		x	[123]
	Centrarchus macropterus	Distinguishing flier	6		x	[3]
	Chaenobryttus coronarius	Bartram	7		x	[36] [3]
	Lepomis auritus	Redbreast sunfish		8	x	[70] [3]
	Lepomis cyanellus	Green sunfish	9		x	[14] [177]
	Lepomis cyanellus	Green sunfish (+)		7.5	x	[70]
	Lepomis gibbosus	Perch		13	x	[71] [3]
	Lepomis gibbosus	Pumpkinseed	12		x	[100]
	Lepomis humilis	Orange-spotted sunfish	4		x	[17] [3]
	Lepomis macrochirus	Bluegill	10		x	[177]
	Lepomis macrochirus		10		x	[108] [3]
	Lepomis macrolophus		8		x	[164] [3]
	Lepomis megalotis	Perch		12	x	[71] [3]
	Micropterus coosae		10		x	[3]
	Micropterus dolomieu		14		x	[71] [3]
	Micropterus punctulatus	Spotted bass	7		x	[188] [3]
	Micropterus punctulatus	Spotted bass	6		x	[151]
	Micropterus salmoides	Largemouth black bass (+)		11	x	[70] [3]
	Micropterus salmoides	Largemouth black bass	16		x	[3]
	Micropterus salmoides	Largemouth bass	15		x	[177]
	Micropterus salmoides	Largemouth black bass (+)		11	x	[70]
	Micropterus salmoides	Largemouth bass	23		x	[82]
	Pomoxis annularis	White crappie	9		x	[135]
	Pomoxis annularis	White crappie	10		x	[177]

	Pomoxis nigromaculatus	Black crappie		12	x	[70]
	Pomoxis nigromaculatus	Black crappie	13		x	[189]
	Pomoxis nigromaculatus	Black crappie	10		x	[177]
	Pomoxis sparoides	Crappie (+)		7.5	x	[70] [3]
Centropomidae	Centropomus undecimalis	Snook	7		x	[3]
	Lates niloticus	Snook		7.8	x	[70] [3]
Chaedontidae	Chaetodon lineolatus	Lined butterflyfish	10		x	[98]
	Chaetodon lunula	Racoon butterflyfish	9		x	[98]
	Chelmon rostratus	Copperband butterflyfish	10		x	[98]
	Forcipiger flavissimus	Long-nosed butterflyfish	18		x	[98]
Chaetodontidae	Centropyge tibicen	Anemonefish	6		x	[142]
Cichlidae	Acara tetramerus			7	x	[70] [3]
	Herichthys cyanoguttatus	Rio Grande cichlid		5	x	[71] [3]
	Tilapia spp.	Cichlid		7	x	[70] [3]
Eleotridae	Eleotris spp.	Sleeper		5.6	x	[70] [3]
Embiotocidae	Amphistichus argenteus	Surfperch	9		x	[38] [3]
	Amphistichus argenteus	Barred surfperch	6		m	[38]
	Amphistichus argenteus	Barred surfperch	9		f	[38]
	Cymatogaster aggregata	Shiner perch	3		x	[19] [3]
	Cymatogaster aggregata	Shiner perch	8		x	[23] [204]
	Cymatogaster aggregata	Shiner perch	3		m	[6]
	Cymatogaster aggregata	Shiner perch	5		f	[6]
	Embiotoca jacksoni	Black perch	10		x	[68]
	Embiotoca lateralis	Surfperch	9		x	[184] [3]
	Hyperprosopon argenteum	Walley surfperch	6		x	[19] [3] [6] [75]
	Hyperprosopon ellipticum	Silver surfperch	5		m	[203]
	Hyperprosopon ellipticum	Silver surfperch	7		f	[203]
	Hysterocarpus traski	Tule perch (+)	5		x	[143]
	Taeniotoca lateralis	Surfperch	9		x	[184] [3]
Gobiidae	Clevelandia ios	Arrow goby	3		x	[153][33]
	Gobius minutus	Goby (+)		2	x	[70] [3]
	Gobius niger	Black goby		5	x	[121]
	Gobius paganellus	Goby		1	x	[70] [3]
	Latrunculus pellucidus			1	x	[70] [3]
	Periophthalmus barbarus	Mudskipper		2.2	x	[154] [3]
	Periophthalmus koelreuteri	Mangrove goby		1.5	x	[70]
Haemulidae	Haemulon sciurus	Grunt	12		x	[98]
Istiophoridae	Istiophorus platypterus	Atlantic sailfish (+)	10		x	[155]
Labridae	Coris julis	Wrasse	7		x	[98]
	Pimelometopon pulchrum	Wrasse	53		x	[3]
	Tautoga onitis	Wrasse		8	x	[70] [3]
Lutjanidae	Neomaenis griseus	Gray snapper		7	x	[70]
	Neomaenis synagris	Spot or lane snapper		7	x	[70]
	Lutjanus griseus	Snapper (+)		7	x	[70] [3]
	Lutjanus peru	Snapper (+)		7.2	x	[154] [3]
	Lutjanus synagris	Snapper (+)		7	x	[70] [3]
Malacanthidae	Caulodatilus princeps	White fish	13		x	[66] [3]
Percichthidae	Macquaria novemaculeata	Australian bass (+)	22		x	[88]
	Morone saxatilis	Striped bass (+)	30		f	[143]
Percidae	Boleosoma longinamum	Darter	4		x	[161] [3]

	Etheostoma blennoides	Appalachian darter	4		x	[114] [3]
	Hadropterus maculatus	Darter		7	x	[70] [3]
	Lucioperca lucioperca	Pikeperch	14		x	[71] [3]
	Lucioperca saudra	Pikeperch		6.9	x	[70]
	Peoilichthys zonalis		3		x	[114] [3]
	Perca flavescens	N. American perch		12	x	[71] [3]
	Perca flavescens	N. American yellow perch (+)		8	x	[70]
	Perca fluviatilis	Perch		5.7	x	[70]
	Percina caprodes	Logperch	3		x	[36] [3]
	Stizostedion canadense	Sauger	18		x	[89]
	Stizostedion canadense	Sauger	13		x	[89] [3]
	Stizostedion vitreum	Walleye	18		x	[3]
	Stizostedion vitreum	Walleye (+)	26		x	[173]
Pholidae	Pholis gunnellus	Gunnel		1.2	x	[70][71] [3]
	Pholis gunnellus	Butterfish	5		x	[159] [3]
Pomacanthidae	Centropyge flavissimus	Lemonpeel angelfish	11		x	[98]
	Euxiphipops navarchus	Blue-faced angelfish	15		x	[98]
	Pomacanthodes imperator	Emperor angelfish	14		x	[98]
	Pygoplites diacantus	Regal angelfish	14.5		x	[98]
	Amphiprion clarkii	Anemonefish	11		x	[142]
	Amphiprion ephippium	Clownfish	16		x	[98]
	Dascyllus aruanus	Damselfish	6		x	[98]
Pomadasyidae	Anisotremus davidsoni	Grunt		7.3	x	[154] [3]
Sciaenidae	Aplodinotus grunniens	Freshwater drum	13		x	[3]
	Atractoscion nobilis	White seabass (+)	20		x	[67]
	Cheilotrema saturnum	Black croaker	20		x	[118] [3]
	Cynoscion macdonaldi	Totoaba	15		x	[25]
	Cynoscion nobilis	White seabass	20		x	[3]
	Cynoscion nothus	Silver seatrout	1.2		x	[56]
	Genyonemus lineatus	White croaker	15		x	[120] [75]
	Genyonemus lineatus	White croaker	15		x	[75]
	Menticirrhus undulatus	Brazilian croaker	8		x	[19] [3]
	Perca fluviatilis	Perch (+)		10.7	x	[71]
	Perca fluviatilis	Perch	10		x	[178]
	Perca fluviatilis	Perch	22		x	[2]
	Pogomias cromis	Drum		6.5	x	[70] [3]
	Roncador stearnsi		15		x	[19] [3]
	Sciaenops ocellata	Redfish drum (+)		7	x	[70] [3]
Scombridae	Germo alalunga	Atlantic albacore	9		x	[180] [3]
	Neothunnus macropterus	Yellowfin tuna	9		x	[180] [3]
	Pneumatophorus diego	Pacific mackerel	9		x	[63] [65]
	Pneumatophorus diego		11		x	[63] [3]
	Scomber scombrus	Mackerel		4	x	[70]
	Scomber scombrus	Mackerel	15		x	[147]
	Scomberomorus maculatus	Spanish mackerel	5		x	[3]
Serranidae	Centropristis striatus	Black seabass	20		x	[28]
	Epinephelus gigas	Mediterranean grouper		24	x	[70][71] [3]
	Lepibema chrysops	White bass	9		x	[181] [3]
	Morone americana		16		x	[3]

	Morone interrupta	Yellow bass	7		x	[117] [3]
	Morone labrax	Bass (+)		10.1	x	[71] [3]
	Paralabrax clathratus	Kelp bass (+)	20		x	[3]
	Paralabrax clathratus	Kelp bass	31		x	[206][61]
	Paralabrax nebulifer	Sand bass	31		x	[131] [195]
	Roccus lineatus	Striped bass or rock-fish		19	x	[70]
	Roccus saxatilus	Bass		24	x	[71] [3]
	Stereolepis gigas		75		x	[3]
Sparidae	Archosargus probatocephalus	'Sheepshead'		6	x	[70] [3]
	Cantharus vulgaris	Sea bream		5	x	[70]
	Spondyliosoma cantharus	Black sea bream		15	x	[70] [3]
	Stenotomus chrysups	Scup	19		x	[28]
Sphyraenidae	Sphyraena argentea	Pacific barracuda	11		x	[197] [3]
Thunnidae	Istiophorus americanus	Sailfish	3.5		x	[54]
	Neothunnus macropterus	Yellowfin tuna	5		x	[139]
	Thunnus thynnus	Bluefin tuna	7		x	[171]
Trichiuridae	Trichiurus lepturus	Ribbon fish	6		x	[3]
Zoarchidae	Macrozoarces americanus	Ocean pout	17		x	[149] [3]
	Macrozoarces americanus	Ocean pout	18		x	[28]
Pleuronectiformes						
Bothidae	Citharichthys sordidus	Lefteye flounder	9		x	[3]
	Citharichthys sordidus	Sanddab	7		m	[10]
	Citharichthys sordidus	Sanddab	8		f	[10]
	Lophopsetta aquosa	Lefteye flounder	7		x	[138] [3]
	Paralichthys californicus	California halibut	15		x	[3]
	Paralichthys californicus	California halibut	30		x	[75] [84] [176] [163]
	Pseudorhombus cinnamoneus	Lefteye flounder	5		x	[124] [3]
	Rhombus maximus	Lefteye flounder	12		x	[3]
Pleuronectidae	Eopsetta jordani	Petrale sole	18		x	[3]
	Glyptocephalus cynoglossus	Witch flounder	14		x	[31] [3]
	Hippoglossus hippoglossus	Atlantic flounder	6		x	[12] [3]
	Hippoglossus hippoglossus	Atlantic flounder	40		x	[71]
	Hippoglossus spp.	Halibut	45		x	[27]
	Hippoglossus stenolepis	Pacific turbot	40		x	[11] [3]
	Hippoglossus vulgaris	Halibut	90		m	[55]
	Hippoglossus vulgaris	Halibut	90		f	[55]
	Hypsopsetta guttulata	Diamond turbot	9		x	[19] [3]
	Isopsetta isolepis	Butter sole	10		m	[91]
	Lepidopsetta bilineata	Rock sole	15		x	[3]
	Limanda aspera	Yellowfin sole	15		x	[3]
	Microstomus pacificus	Dover sole	15		x	[85] [3]
	Microstomus pacificus	Sole	45		x	[27]
	Platichthys stellatus	Starry flounder	13		x	[3]
	Platichthys stellatus	Starry flounder	17		f	[34]
	Platichthys stellatus	Starry flounder	24		m	[34]
	Pleuronectes platesca	Plaice	13		m	[26]
	Pleuronectes platesca	Plaice	22		f	[26]
	Pleuronectes platesca	Plaice (+)	30		x	[71]
	Pleuronectes platesca	Plaice	45		x	[27]

	Pleuronectes platesca	Plaice		2.9	x	[70]
	Pseudopleuronectes americanus	Winter flounder	10		x	[57]
	Pseudopleuronectes americanus	Winter flounder		1	x	[70]
Soleidae	Solea solea	Common sole	20		x	[27]
	Solea solea	Common sole	9		x	[42] [3]
	Solea vulgaris	Dover sole	8		x	[28]
Polypteriformes						
Polypteridae	Polypterus senegalus			34	x	[71] [3]
	Polypterus senegalus			5.7	x	[70]
Salmoniformes						
Coregonidae	Coregonus clupeaformis	Lake whitefish	26		x	[90] [3]
	Coregonus clupeaformis	Whitefish	27		x	[90]
	Coregonus clupeaformis	Lake whitefish		12	x	[71]
	Coregonus clupeaformis	Lake whitefish	28		x	[110]
	Coregonus spp.	Whitefish (+)	30		x	[27]
	Cristivomer namaycush	Lake trout	18.4		x	[111]
	Hypomesus olidus	Pond smelt	10		x	[170]
	Leucichthys artedi	Cisco	11		x	[96]
	Leucichthys artedii		13		x	[90] [3]
	Leucichthys nigripinnis	Cisco	11		x	[108] [3]
	Leucichthys sardinella	Whitefish	11		x	[202][201]
	Leucichthys zenithicus	Long jaw	10		x	[196] [3]
	Lovellia scali	Tasmanian whitebait	2		x	[29]
	Oncorhynchus clarki	Cutthroat trout	10		x	[190] [177] [104]
	Oncorhynchus clarkii	Cutthroat trout	10		x	[107]
	Oncorhynchus gorbuscha	Pink salmon	1.8		x	[3]
	Oncorhynchus gorbuscha	Pink salmon	3		x	[177]
	Oncorhynchus gorbuscha	Pink salmon	0.1		x	[93]
	Oncorhynchus keta	Chum salmon	5		x	[86] [3]
	Oncorhynchus keta	Chum salmon	7		x	[177]
	Oncorhynchus keta	Chum salmon	0.6		x	[93]
	Oncorhynchus keta	Chura salmon	6		x	[122]
	Oncorhynchus kisutch	Coho salmon	4		x	[3]
	Oncorhynchus kisutch	Coho salmon	5		x	[177] [143]
	Oncorhynchus kisutch	Coho salmon	4.1		x	[93]
	Oncorhynchus mykiss	Steelhead	9		x	[199]
	Oncorhynchus mykiss	Steelhead	7		x	[107]
	Oncorhynchus nerka	Sockeye salmon	8		x	[16] [3] [73]
	Oncorhynchus nerka	Sockeye salmon	6		x	[72]
	Oncorhynchus nerka	Sockeye salmon	4.3		x	[93]
	Oncorhynchus tshawytscha	Chinook salmon	7		x	[3]
	Oncorhynchus tshawytscha	Chinook salmon	9		x	[143] [61]
	Oncorhynchus tshawytscha	Chinook salmon	5		x	[91]
	Oncorhynchus tshawytscha	Chinook salmon	2.5		x	[93]
	Prosopium cylindraceum		14		x	[162] [3]
	Prosopium williamsoni		17		x	[3]
	Salmo clarki		10		x	[8] [3]
	Salmo fario	Brown trout		6.8	x	[70]
	Salmo gairdneri	Steelhead trout (+)		3.9	x	[70] [3]

	Salmo gairdneri	Steelhead trout	9		x	[3]
	Salmo gairdneri	Steelhead trout	8		x	[177]
	Salmo lacustris	Lake trout		10.3	x	[70]
	Salmo salar	Atlantic salmon	13		x	[71]
	Salmo salar	Atlantic salmon		3.3	x	[70]
	Salmo salar	Atlantic salmon	5		x	[146]
	Salmo salar salar	Atlantic salmon	13		x	[71] [3]
	Salmo salar sebago	Atlantic salmon	8		x	[36] [3]
	Salmo trutta	Brown trout		10	x	[71]
	Salmo trutta	Brown trout	18		x	[146]
	Salmo trutta	Brown trout	8		x	[78]
	Salmo trutta	Brown trout	11		m/f	[106]
	Salvelinus alpinus	Char	22		x	[186] [3]
	Salvelinus alpinus	Char	24		x	[80]
	Salvelinus aureolus		5		x	[3]
	Salvelinus fontinalis	Splake	8		x	[79] [3]
	Salvelinus fontinalis	Splake	16		m	[74]
	Salvelinus fontinalis	Splake	16		f	[74]
	Salvelinus fontinalis	Bunnylake brook trout	24		x	[166]
	Salvelinus malma	Char		20	x	[71] [3]
	Salvelinus namaycush	Char		12	x	[71]
	Salvelinus namaycush	Char	41		x	[186]
	Salvelinus spp.	Char (+)	30		x	[27]
	Salvelinus trutta	Brook trout	8		x	[107]
	Stenodus leucichthys		16		x	[58] [3]
Hiodontidae	Amphiodon alosoides		10		x	[135] [3]
	Hiodon tergisus		7		x	[15] [3]
Osmeridae	Hypomesus pretiosus	Surf smelt	5		x	[205]
	Hypomesus pretiosus	Surf smelt	3		x	[205]
	Hypomesus transpacificus	Delta smelt	1		x	[60] [143]
	Mallotus villosus	Capelin	7		x	[3]
	Mallotus villosus	Capelin	10		x	[27]
	Osmerus mordax	Smelt	6		x	[126] [3]
	Osmerus mordax	Smelt	7		x	[13]
	Spirinchus thaleichthys	Longfin smelt	2		x	[141]
	Spirinchus thaleichthys	Longfin smelt (+)	2		f	[143]
	Thaleichthys pacificus	Columbia river smelt	4		x	[185] [3]
	Thaleichthys pacificus	Eulachon	5		x	[185]
Salmonidae	Argentina semifaxiatn	Argentine	2		x	[87]
Thymallidae	Thymallus signifer		11		x	[134] [3]
	Thymallus vulgaris	Grayling		2.8	x	[70]
Scomberesociformes						
Scomberesocidae	Cololabis saira		4		x	[137] [3]
Scorpaeniformes						
Anoplopomatidae	Anoplopoma fimbrim	Sablefish	20		x	[3]
Cottidae	Cottus burbalis	Sculpin		3.4	x	[71] [3]
	Leptocottidae armatus	Pacific staghorn sculpin	10		x	[204]
	Myoxocephalus octodecemspinosus	Longhorn sculpin	11		x	[140] [3]
Hexagrammidae	Ophiodon elongatus	Lingcod	20		x	[130]
	Ophiodon elongatus	Lingcod	16		x	[3]

Platycephalidae	Neoplatycephalus macrodon	Australian trawl fish	6		x	[3]
Scorpaenidae	Medialuna californiensis	Halfmoon cleaner fish	8		x	[19] [3]
	Pontinus clemensi	Scorpionfish	8		x	[64] [3]
	Pterois volitans	Scorpionfish	10		x	[98]
	Sebastes aleutianus	Rockfish	140		x	[41]
	Sebastes alutus	Pacific ocean perch	26		m	[22]
	Sebastes alutus	Pacific ocean perch	28		f	[22]
	Sebastodes alutus	Rockfish	20		x	[4] [3]
	Sebastes diploproa	Splitnose rockfish	40		x	[24]
	Sebastes marinus	Redfish	50		x	[109] [3]
	Sebastes mentella	Redfish	75		x	[35]
	Sebastes mentella	Rockfish	50		x	[169]
	Sebastes spp.	Ocean perch (+)	80		x	[27]
Triglidae	Trigla hirundo	Sea robin		6.8	x	[70][71] [3]
	Trigla lucerna	Yellow gurnard	4		m	[150]
	Trigla lucerna	Yellow gurnard	10		f	[150]
Lepisosteiformes						
Lepisosteidae	Lepisosteus osseus	Longnosed gar	30		x	[16] [3]
	Lepisosteus osseus	Longnosed gar		24	x	[71]
	Lepisosteus osseus	Gar	30		x	[98]
	Lepisosteus osseus	Longnosed gar		20	x	[70]
	Lepisosteus platostomus	Shortnosed gar (+)		20	x	[70] [3]
Siluriformes						
Ameiuridae	Ictalurus lacustris punctatus	Channel catfish	12		x	[9]
Ariidae	Galeichthys felis	Catfish (+)		7	x	[70] [3]
Clariidae	Clarias lazera	African catfish (+)		16.2	x	[70] [3]
Ictaluridae	Ameiurus melas		9		x	[3]
	Ameiurus natalis		4		x	[116] [3]
	Ameiurus nebulosis		3		x	[116] [3]
	Ictalurus catus	Catfish (+)		8.1	x	[70]
	Ictalurus catus	Catfish	14		x	[175]
	Ictalurus lacustris	Channel catfish	13		x	[116] [3]
	Ictalurus melas	Black bullhead (+)	10		x	[151]
	Pilodictis olivaris		15		x	[18] [3]
Loricariidae	Plecostomus punctatus	Armored catfish	18		x	[98]
Mochokidae	Synodontis schall	Upside-down catfish		31	x	[70][71] [3]
	Synodontis schall	Upside-down catfish		9	x	[70]
	Synodontis schall	Upside-down catfish	12		x	[98]
Pimelodidae	Pimelodus spp.	Catfish/pirana		7	x	[70] [3]
Siluridae	Amiurus catus	N. American catfish		8.1	x	[70]
	Bagrus bayad	African catfish		17	x	[70]
	Callichthys punctatus	S. American catfish		8.8	x	[70]
	Chrysichthys auratus	African catfish		5.8	x	[70]
	Hexanematichthys felis	N. American sea-catfish (+)		7	x	[70]
	Malopterurus electricus	African catfish		7.1	x	[70]
	Saccobranchus fossilis	Asiatic catfish (+)		17.5	x	[70] [3]
	Schilbe mystus	African catfish		9.7	x	[70]
Tetraodontiformes						
Tetraodontidae	Tetraodon fahaka			5.9	x	[70] [3]
	Tetraodon fahaka	Nile globe-fish		5.9	x	[70]

Tetraodon solandri	Puffer	7	x	[71] [3]

References

1. Ackerman, L.T., 1971. *Contributions to the Biology of the Leopard Shark, Triakis semifasciata (Girard) in Elkhorn Slough, Monterey Bay, California.* Sacramento, CA: Sacramento State College.
2. Alm, G., 1951. Year class fluctuations and span of life of perch. Drottningholm: Institute of Freshwater Research, Drottningholm, Fishery Board of Sweden.
3. Altman, P.L. and D.S. Dittmer, eds. 1962. *Growth, Including Reproduction and Morphological Development.* Biological Handbooks. Washington, D. C.: Federation of American Societies for Experimental Biology.
4. Alverson, D. and S. Westrheim, 1961. *Rapports Proces-Verbaux Reunions Conseil Permanent Intern Exploration Mer,* Vol. 150: 12.
5. Anderson, A.W., 1954. 152-Year old lake sturgeon caught in Ontario. *Commercial Fisheries Review* 18: 28.
6. Anderson, R.D. and C.F. Bryan, 1970. Age and growth of three surfperches (Embioticidae) from Huboldt Bay, California. *Transactions of American Fisheries Society* 99: 475-482.
7. Anderson, R.S., 1988. Columbia River Sturgeon. Seattle, WA: Washington Sea Grant Report No. WSG AS88-14.
8. Anonymous, 1951. *Annual Report Fishery Division.* State Game Commission Oregon.
9. Applegate, V.C. and B.R. Smith, 1950. Sea lamprey spawning runs in the Great Lakes in 1950. *U.S. Fish and Wildlife Service Scientific Report: Fisheries* Vol. 61.
10. Arora, H.L., 1951. An investigation of the California sand dab, *Citharichthys sordidus* (Girard). *California Fish and Game* 37: 1-42.
11. Babcock, J., 1929. Canadian fisherman *Canadian Fisherman,* Vol. 16: 39.
12. Bagenal, T., 1955. The growth rate of the rough dab *Hippoglosssoides platessoides* (Fabr.). *Journal of Marine Biological Association* 34: 297.
13. Bailey, M.M., 1964. Age, growth, maturity, and sex composition of the American smelt, *Osmerus mordax* (Mitchill) of Western Lake Superior. *Transactions of American Fisheries Society* 93: 382.
14. Bailey, R.M. and K.F. Lagler, 1938. An analysis of hybridization in a population of stunted sunfishes in New York. *Michigan Academy of Sciences* 23: 588.
15. Bajkov, A., 1930. [No title]. *Transactions of American Fisheries Society* 60: 215.
16. Barnaby, J.T., 1944. Fluctuations in abundance of red salmon, *Oncorhynchus nerka* (Walbrim) of the Karluk River, Alaska. *U. S. Fish Wildlife Service Fishery Bulletin* 50: 237.
17. Barney, R.L. and B.J. Anson, 1923. Life history and ecology of the orange spotted sunfish, *Lepomis humilis. U. S. Fisheries Commisson Report,* Vol. 15.
18. Barnickol, P.G. and W.C. Starrett, 1951. Commercial and sport fisheries of the Mississippi River between Caruthersville, Missouri, and Dubuque, Iowa. *Illinois Natural History Survey Bulletin,* Vol. 25: 267.
19. Baxter, J.L., 1960. *Inshore Fishes of California.* Sacramento: California Department of Fish and Game.

20. Baxter, J.L., 1960. A study of the yellowtail *Seriola dorsalis* (Gill). *California Department of Fish and Game Bulletin,* Vol. 110: 29.

21. Baxter, J.L., 1967. Summary of biological information on the northern anchovy *Engraulis mordax girard. California Cooperative Oceanic Fisheries Investigative Report,* Vol. 11: 110-116.

22. Beamish, R.J., 1979. New information on the longevity of Pacific ocean perch (*Sebastes alutus*). *Journal of the Fisheries Research Board of Canada* 36: 1395-1400.

23. Beardsley, A.J. and C.E. Bond, 1970. Field Guide to Common Marine and Bay Fishes of Oregon. *Agricultural Experiment Station Bulletin,* Vol. 607: 27. Corvallis: Oregon State University.

24. Bennett, J.T., G.W. Boehlert, and K.K. Turekian, 1982. Confirmation of longevity in *Sebastes diploproa* (Pisces: Scopraenidae). *Marine Biology* 71: 209-215.

25. Berdegue, J.A., 1955. La pesqueria de la totoaba (*Cynoscion macdonaldi* Gilbert) en San Felipe, Baja, California. *Revista De La Sociedad Mexicana De Historia Natural* 16: 45-78.

26. Beverton, R.J.H., 1982. Unpublished.

27. Beverton, R.J.H., J.R. Beddington, and D.M. Lavigne, eds. 1985. *Marine Mammals and Fisheries*. Boston: G. Allen & Unwin.

28. Beverton, R.J.H. and S.J. Holt, 1959. A review of the lifespans & mortality rates of fish in nature and their relationship to growth and other physiological characteristics. *Ciba Foundation Colloquia on Ageing,* Vol. 5: 142-180.

29. Blackburn, M., 1950. The Tasmanian whitebait , *Lovettia seali* (Johnston) and the whitebait fishery. *Australian Journal of Marine and Freshwater Research* 1: 155-198.

30. Bobick, J.E. and M. Peffer, eds. 1993. *Science and Technology Desk Reference*. Washington, D.C.: Gale Research Inc.

31. Bowers, A.B., 1960. Growth of the witch (*Glyptocephalus cynoglossus*) in the Irish Sea. *Conseil Permanent Interior Exploration Mer,* Vol. 25.

32. Breder, C.M.J., 1936. *Bulletin New York Zoological Society* 39: 116.

33. Brothers, E.B., 1975. *The Comparative Ecology and Behavior of Three Sympatric California Gobies.* San Diego, CA: University of California, San Diego.

34. Campana, S.E., 1984. Comparison of age determination methods for the starry flounder. *Transactions of the American Fisheries Society* 113: 365-369.

35. Campana, S.E., 1990. Determination of longevity in redfish. *Canadian Journal of Fisheries and Aquatic Sciences* 47: 163-165.

36. Carlander, K.D., 1950. *Handbook of Freshwater Fishery Biology*. Dubuque, IA: W. C. Brown.

37. Carlander, K.D., 1969. *Handbook of Freshwater Biology I. Life History data on Freshwater fishes of the United States and Canada, exclusive of the Perciformes* Vol. 1. Ames: Iowa State University Press.

38. Carlisle, J.G.J., J.W. Scott, and N.J. Abramson, 1960. The barred surfperch (*Amphistichus argentas agassiz*) in southern California. *California Department of Fish and Game Bulletin,* Vol. 68: 443.

39. Caywood, M.L., 1974. *Contributions to the Life History of the Split-tail Pogonichthys macrolepidotus (Ayres).* Sacramento: California State University, Sacramento.

40. Chang, H.-W., 1951. Age and growth of *Callionymus lyra* L. *Journal of the Marine Biological Association of the United Kingdom* 30: 281-295.

41. Chilton, D.E., 1982. Age determination methods for fishes studies by the groundfish program at the Pacific biological station. In D.E. Chilton and R.J. Beamish (eds.), *Canadian Special Publication of Fisheries and Aquatic Sciences,* Vol. 102. Ottawa: Department of Fisheries and Oceans.

42. Christensen, J., 1960. The stock of soles (*Solea solea*) and the sole fishery on the Danish North Sea coast. *Danmarks Fisk--Havundersokelser Medd.* 3: 19.

43. Clark, F.N., 1925. The life history of *Leuresthes tenuis*an atherine fish with tide controlled spawning habits. *California Fish and Game Fish Bulletin* 10: 51.

44. Clark, F.N., 1929. The life history of the California jack smelt, *Atherinopsis californienses. California Fish and Game* 16: 22.

45. Clark, F.N., 1938. Grunion in Southern California. *California Department of Fish and Game Bulletin,* Vol. 24: 49.

46. Clark, F.N., 1940. The application of sardine life history to the industry. *California Fish and Game Bulletin* 26: 39-48.

47. Clark, F.N. and J.B. Phillips, 1952. The northern anchovy (*Engraulis mordax*) in California fishery. *California Fish and Game* 38: 198.

48. Clark, J., 1963. Aquarium Records.

49. Clemens, W.A., 1939. The fishes of Okanagan Lake and nearby waters. *Fisheries Research Board of Canada Bulletin* 56: 27.

50. Clemmens, W.A. and G.V. Wilby, 1961. Fishes of the Pacific coast of Canada. *Fisheries Research Board of Canada* 68: 443.

51. Compagno, L.J.V., 1988. *Sharks of the Order Carcharhiniformes*. Princeton: Princeton University Press.

52. Cooper, G.P., 1936. Papers of Michigan Academy of Science. 22: 587.

53. Crow, G.L. and J.D. Hewitt, 1988. Longevity records for captive Tiger sharks *Galeocerdo Cuvier* with notes on behavior management. *International Zoological Yearbook,* Vol. 27: 237-240.

54. DeSylva, D.P., 1957. Studies on the age and growth of the Atlantic sailfish, *Istiophorus americanus* (Cuvier), using length-frequency curves. *Bulletin of Marine Science of the Gulf and Caribbean* 7: 1-20.

55. Devold, F., 1938. The north Atlantic halibut and net fishing. *Fiskeridirektoratet Reports on Norwegian Fishery and Marine Investigations,* Vol. 5.

56. deVries, D.A. and M.E.J. Chittenden, 1982. Spawning, age determination, longevity, and mortality of the silver seatrout, *Cynoscion nothus* in the Gulf of Mexico. *U.S. Fish and Wildlife Service Fishery Bulletin* 80: 487-500.

57. Dickie, L.M. and F.D. McCracken, 1955. Isopleth diagrams to predict equilibrium yields of a small flounder fishery. *Journal of the Fisheries Research Board* 12: 187-209.

58. Dymond, J.R., 1943. The coregonine fishes of northwestern Canada. *Transactions of the Royal Society of Canada* 24: 172.

59. Emmett, R.L., 1991. *Distribution and Abundance of Fishes and Invertebrates in West Coast Estuaries. II: Species Life History Summaries*. Rockville, MD: U.S. Dept. of Commerce; National Oceanic and Atmospheric Administration, National Ocean Service.

60. Erkkila, L.F., J.W. Moffett, O.B. Cope, B.R. Smith, and R.S. Nielson, 1950. Sacramento-San Joaquin Delta fishery resources: effects of Tracy pumping plant and delta cross channel. *U.S. Fish and Wildlife Servie Special Scientific Report* 56: 109.

61. Eschmeyer, W.N., W.S. Herald, and H. Hammann, 1983. *A Field Guide to Pacific Coast Fishes of North America*. Boston, MA: Houghton Mifflin Co.

62. Feder, H.M., C.H. Turner, and C. Limbaugh, 1974. Observations of fishes associated with kelp beds in Southern California. *California Department of Fish and Game Bulletin* 160: 144.

63. Fitch, J.E., 1951. Age composition of the southern catch of Pacific mackerel 1939-1940 and 1950-1951. *California Fish and Game* 83: 27.

64. Fitch, J.E., 1955. *Pontinus clemensi*, a new scorpaenoid fish from the tropical eastern Pacific. *Journal Washington Academy of Science* 45: 61.

65. Fitch, J.E., 1956. Age composition of the southern California catch of Pacific mackerel for the 1954-55 season. *California Fish and Game* 42: 143-148.

66. Fitch, J.E., 1960. *Offshore Fishes of California*. Sacramento, CA: California Department of Fish and Game.

67. Fitch, J.E. and R.J. Lavenberg, 1971. *Marine Food and Game Fishes of California. California Natural History Guides 28*. Berkeley, CA: University of California.

68. Fitch, J.E. and R.J. Lavenberg, 1975. *Tidepool and Nearshore Fishes of California* Vol. 38. Berkeley, CA: University California Press.

69. Fleming, A.M., 1960. Age, growth and sexual maturity of Cod (*Gadius morhua* L.) in the Newfoundland Area, 1947 - 1950. *Journal of the Fisheries Research Board of Canada* 17: 775.

70. Flower, M.S.S., 1925. Contributions to our knowledge of the duration of life in vertebrate animals - I. Fishes. *Proceedings of the Zoological Society of London* 1925(1): 247-267.

71. Flower, M.S.S., 1935. Further notes on the duration of life in animals. - I. Fishes: as determined by otolith and scale - readings and direct observations on living animals. *Proceedings of the Zoological Society of London* 1935: 265.

72. Foerster, R.E., 1929. An investigation of the life history and propagation of the sockeye salmon (*Oncorhynchus nerka*) at Cultus Lake, British Columbia. No. 1. Introduction and the run of 1925. *Contributions to Canadian Biology and Fisheries* 5: 1-35.

73. Foerster, R.E., 1968. The sockeye salmon, *Onchorhynchus nerka. Fish Research Board Canada* Bulletin No. 162: 422.

74. Fraser, J.M., 1983. Longevity of first generation splake. *Progressive Fish-Culturist* 45: 233-234.

75. Frey, H.W., 1971. *California's Living Marine Resources and Their Utilization*. Sacramento, CA: California Department Fish Game.

76. Frost, W.E., 1943. The natural history of the minnow, *Phoxinus phoxinus*. *Journal of Animal Ecology* 12: 139-162.
77. Frost, W.E., 1945. The age and growth of eels (*Anguilla anguilla*) from the Windermere catchment area. *Journal of Animal Ecology* 14: 106-124.
78. Frost, W.E. and W.J.P. Smyly, 1952. The brown trout of a morrland fish pond. *Journal of Animal Ecology* 21: 62-86.
79. Fry, H.D.J., 1936. Life history of *Hesperoleucus venustus* Snyder. *California Fish and Game* 22: 65.
80. Grainger, E.H., 1953. On the age, growth, migration, reproductive potential and feeding habits of the Arctic char (*Salvelinus alpinus*) of Frobisher Bay, Baffin Island. *Journal of the Fisheries Research Board of Canada* 10: 326-371.
81. Grant, C.J. and et.al., 1979. Estimation of growth, mortality and yield per recruit of the Australian school shark, *Galeorhinus australis* (Macleay) from tag recoveries. *Australian Journal of Marine and Freshwater Research* 30 (5): 625-637.
82. Green, D.M. and R.C. Heidinger, 1994. Longevity record for largemouth bass. *North American Journal of Fisheries Management* 14: 464-465.
83. Greenbank, J. and P.R. Nelson, 1959. Life history of the threespine stickleback *Gasterosteus aculeatus linnaeus* in Karbuk Lake and Bore Lake, Kodiak Island, Alaska. *U. S. Fish Wildlife Service Fishery Bulletin* 153: 537.
84. Haaker, P.L., 1975. *The biology of the California halibut, Paralichthys californicus (Ayers), in Anaheim Bay, California* Vol. Fish Bull. 165: California Fish Game.
85. Hagerman, F.B., 1952. The biology of the dover sole, *Microstomus pacificus* (Lockington). *California Division of Fish and Game Bulletin* 85: 31.
86. Hammett, F.S. and D.W. Hammett, 1939. Proportional length growth of gar (*Lepisoseus platyrhincus* de Kay). *Growth* 3: 197.
87. Hanyu, I., 1956. On the age and growth of argentine (*Argentina semifasciata* Kishinouye). *Japanese Society of Scientific Fisheries* 21: 991-999.
88. Harris, J.H., 1985. Age of Australian bass, *Macquaria novemaculeata* (Perciformes: Percichthyidae), in the Sydney Basin. *Australian Journal of Marine and Freshwater Research* 36: 235-246.
89. Hart, J.L., 1928. Data on the rate of growth of pike perch (*Stizostedion vitreum*) and sauger (*S. canadense*) in Ontario. *University of Toronto, Ontario Fisheries Research Laboratory* 34: 44-55.
90. Hart, J.L., 1931. The growth of the whitefish *Coregonus clupeaformis* (Mitchill). *Contributions to Canadian Biology and Fisheries* 6: 427-444.
91. Hart, J.L., 1973. Pacific fishes of Canada. *Fisheries Research Board Canada Bulletin* Vol. 180: 740.
92. Hart, T., 1950. A note on growth zones in bones of *Phycis blennoides brunn*. *Journal Conseil Permanent International Exploration Mer* 16: 335.
93. Healey, M.C., 1986. Salmonid age at maturity. *Canadian Special Publication of Fisheries and Aquatic Sciences* 89: 39-52.

94. Heath, K.L., 1980. *Comparative Life Histories of 2 Species of Anchovies, Anchoa delicatissima and A. compressa (*F. Engraulidae*) from Newport Bay, CA.* Fullerton, CA: California State University.

95. Herald, E.S. and M. Rakowicz, 1951. Stable requirements for raising sea horses. *Aquarium Journal* 22: 234-242.

96. Hile, R., 1931. *Indiana Lakes Streams Invest.* 2: 9.

97. Hill, C.M., 1985. Klamath River Basin Fisheries Resource Plan: Various pagination: U. S. Department of Interior.

98. Hinton, 1962. Horned shark, gar, mormyriad, characin, carp, armored catfish, arowana, upside down catfish.

99. Hoenig, 1979. Bull shark, dusky shark.

100. Hubbell, P.M., 1966. Pumpkinseed sunfish, in *Inland Fisheries Management*, Calhoun, Editor.: California Dept. Fish. Game. 402-404, 405-407.

101. Hubbs, C.L., 1921. An ecological study of the life-history of the fresh-water atherine fish *Labidesthes sicculus. Ecology* 2: 262-276.

102. Jester, D.B., 1982. Age and growth, fecundity, abundance and biomass production of the white sands pupfish, Cyprinodon tularosa (Cyprinodontidae) in a desert pond. *Southwestern Naturalist* 27: 43-54.

103. Johnson, J.E., 1970. *Age, Growth, and Population Dynamics of the Threadfin Shad Dorosoma petenense (Gunther), in Central Arizona Resevoirs.* Tempe, AZ: Arizona State University, Tempe.

104. Jones, D.E., 1977. Life history of steelhead trout and life history of sea-run cutthroat trout. *Alaska Department Fish Game* 18: 52-105.

105. Jones, J.W. and H.B.N. Hynes, 1950. The age and growth of *Gasterosteus aculeatus Pygosteus pungitius* and *Spinachia vulgaris*, as shown by their otoliths. *Journal of Animal Ecology* 19: 59-73.

106. Jonsson, B., J. Henning, L'Abee-Lund, T.G. Heggberget, A.J. Jensen, B.O. Johnsen, T.F. Naesje, and L.M. Saettem, 1991. Longevity, body size, and growth in anadromous brown trout *Salmo strutta. Canadian Journal of Fisheries and Aquatic Science* 48: 1838-1845.

107. Kearn, G.C., 1990. The rate of development and longevity of the monogean skin parasite *Entobdella soleae. Journal of Helminthology* 64: 340-342.

108. Keleher, J.J., 1952. Growth and trianenophorus parasitism in relation to taxonomy of Lake Winnipeg ciscoes (*Leucichthys*). *Journal of Fisheries Research Board Canada* 8: 469.

109. Kelly, G.F. and R.S. Wolf, 1959. Age and growth of the redfish (*Sebastes marinus*) in the Gulf of Maine. *U. S. Fish Wildlife Service Fishery Bulletin* 156: 1.

110. Kennedy, W.A., 1953. Growth, maturity, and mortality in the relatively unexploited Lake Trout, *Cristiromer Namaycush*, of Great Slave Lake. *Journal of Fisheries Research Board Canada* 11: 827.

111. Kennedy, W.A., 1954. Growth, maturity and mortality in the relatively unexploited Lake Trout *Cristivomer namaycush*, of Great Slave Lake. *Journal of the Fisheries Research Board of Canada* 11: 827-852.

112. Krumholtz, L.A., 1948. Reproduction in the western mosquitofish, *Gambusia affinis affinis* (Baird and Girard), and its use in mosquito control. *Ecology Monograph* 18: 1-43.

113. Kulkarni, C.V., 1983. Longevity of fish *Megalops cyprinoides* (Broiss). *Journal of the Bombay Natural History Society* 80: 230-232.

114. Lachner, E.A., E.F. Westlake, and P.S. Handwerk, 1950. Studies on the biology of some percid fishes from western Pennsylvania. *American Midland Naturalist* 43: 92.

115. Lea, B.E., 1930. Mortality in the tribe of Norwegian herring. *Rapports et Proces-Verbaux Des Reunions* 65: 100-117.

116. Lewis, W.M., 1950. Fisheries investigations on two artificial lakes in southern Iowa II. Fish populations. *Iowa State College Journal of Science* 24: 287.

117. Lewis, W.M. and K.D. Carlander, 1948. Growth of the yellow bass, *Morone interrrupta gill*, in Clear Lake, Iowa. *Iowa State College Journal of Science* 22: 185.

118. Limbaugh, C., 1961. Life history and ecologic notes on the black croaker. *California Fish and Game* 47: 163.

119. Ling, J.K., 1958. The sea garfish, *Reporhamphus melanochir* (Cuvier and Valenciennes) (Hemiramphidae) in South Australia: breeding, age determination, and growth rate. *Australian Journal of Marine and Freshwater Research* 9: 60.

120. Love, M.S., G.E. McGowen, W. Westphal, R.J. Lavenberg, and L. Martin, 1984. Aspects of the life history and fishery of the white croaker, *Genyonemus lineatus* (Sciaenidae), off California. *Fish Bulletin, U. S.* 82: 179-198.

121. Magnhagen, C., 1990. Reproduction under predation risk in the sand goby, *Pomatoschistus minutus* and the black goby, *Gobius niger:* The effect of age and longevity. *Behavioral Ecology and Sociobiology* 26: 331-335.

122. Marr, J.C., 1943. Age, length, and weight studies of three species of Columbia River salmon (*Oncorhynchus keta, O. gorbuscha*, and *O. kisutch). Stanford Ichthyological Bulletin* 2: 157-197.

123. Mathews, S.B., 1962. *The Ecology of the Sacramento Perch, Archoplites interruptus from Selected Areas of California and Nevada.* Berkeley, CA: University of California, Berkeley.

124. Matsuura, S., 1961. Age and growth of flatfish, ganzo birame, *Pseudorhombus cinnamoneus* (Temmick et Shleger). *Record Oceanographic Works Japan Special Report* 5: 103.

125. McCrimmon, H.R., 1968. Carp in Canada. *Bulletin of the Fisheries Reserve Board of Canada* 165: 93.

126. McKenzie, R.A., 1958. Age and growth of smelt, *Osmerus mordax* (Mitchell) of the Miramichi River, New Brunswick. *Journal of Fisheries Research Board Canada* 15: 1313.

127. Menon, M.D., 1951. Biomedics of the poor cod (*Gadus minutus L.*) in the Plymouth area. *Journal of the Marine Biological Association of the United Kingdom* 29: 185-239.

128. Mense, J.B., 1967. Ecology of the Mississippi silversides, *Menidia audens hay*, in Lake Texoma. *Oklahoma Fisheries Research Laboratory Bulletin* 6: 32.

129. Messtorff, J., 1959. Unterushugen Uber Die Biologie des Whittlings *Merlangius merlangus* (L.) in der Nordsee. *Ber. deut. wiss. Komm. Meeresforsch., n. F.* 15: 277.

130. Miller, D.J. and J.J. Geibel, 1973. Summary of blue rockfish and lingcod life histories; A reef ecology study; and giant kelp, *Macrocystis pyrifera*, experiments in Monterey Bay, California. *California Fish Game* 158:137.

131. Miller, D.J. and R.N. Lea, 1972. *Guide to the coastal marine fishes of California*. California Fish and Game Vol. 157.

132. Miller, D.J. and J. Schmidtke, 1956. Report on the distribution and abundance of Pacific herring (*Clupea pallasi*) along the coast of central and southern California. *California Division of Fish and Game Bulletin* 42: 163-187.

133. Miller, E.V. and W.A. Kennedy, 1948. Observations on the lake trout of Great Bear Lake. *Journal of Fisheries Research Board Canada* 7: 176.

134. Miller, R.B., 1946. Notes on the Arctic grayling, *Thymallus signifer* Richardson, from Great Bear Lake. *Copeia* : 227.

135. Miller, R.B., 1949. *Preliminary Biological Surveys of Alberta Watershed, 1947-1949*. Edmonton, Alberta: Alberta Dep. of Lands and Forests.

136. Mills, C.A., 1983. The age, growth and reproduction of the stone loach *Noemacheilus barbatulus* (L.) in a Dorset chalk stream. *Freshwater Biology* 13: 283-292.

137. Miyauti, T., 1936. On the bionomics of *Cololabis sairain* the north-eastern region of Japan. *Bulletin of the Japanese Society of Science Fisheries* 5: 372.

138. Moore, E., 1947. Studies on the marine resources of southern New England VI. The sand flounder, *Lophopsetta aquosa* (Mitchill): a general study of the species with special emphasis on age determination by means of scales and otoliths. *Bulletin of the Bingham Oceanography Collection* 11: 1.

139. Moore, H.L., 1951. Estimation of age and growth of yellowfin tuna (*Neothunnus macropterus*) in Hawaiian waters by size frequencies. *Fishery Bulletin 65,* Vol. 52: 133-149. Washington D. C.: U. S. Government Printing Office.

140. Morrow, J., 1951. Studies on the marine resources of southern New England: VIII. The biology of the longhorn sculpin *Myoxocephalus octodecimspinosus* Mitchell, with a discussion of the southern New England "trash" fishery. *Bulletin of the Bingham Oceanography Collection* 13: 1.

141. Moulton, L.L., 1974. Abundance, growth, and spawning of the longfin smelt in Lake Washington. *Transactions American Fisheries Society* 103: 46-52.

142. Moyer, J.T., 1986. Longevity of the anemonefish *Amphiprion clarkii* at Miyake-jima, Japan with notes on four other species. *Copeia* 1: 135-139.

143. Moyle, P.B., 1976. *Inland Fishes of California*. Berkeley, CA: University California Press.

144. Murphy, G.I., 1948. Notes on the biology of the Sacramento hitch (*Lavinia exilicauda e.*) of Clear Lake, Lake county, California. *California Fish and Game* 39: 101.

145. Murphy, G.I., 1950. The life history of the greaser blackfish (*Orthodon microlepidotus*) of Clear Lake, Lake County, California. *California Fish and Game* 36: 119.

146. Nall, G.H., 1930. *The Life of Sea Trout.* London: Seeley Service.

147. Nedelec, C., 1958. Biologie et peche du Maquereau. *Revue Trav Institute Pecheur Maritime* 22: 121.

148. O'Connell, C.P., 1953. The life history of the cabezon *Scorpaenichthys marmoratus* (Ayres). *California Division of Fish and Game Bulletin* 93: 60.

149. Olsen, Y.H. and D. Merriman, 1946. Studies on the marine resources of southern New England IV. The biology and economic importance of the ocean pout, *Macrozoarces americanus* (Bloch and Schneider). *Bulletin of the Bingham Oceanography Collection* 9: 1.

150. Papaconstantinou, C., 1984. Age and growth of the yellow gurnard (*Trigla lucerna L.* 1758) from the Thermaikos Gulf (Greece) with some comments on its biology. *Fisheries Research (Amsterdam)* 2: 243-255.

151. Pflieger, W.L., 1975. *The Fishes of Missouri*. Missouri Department of Conservation.

152. Phillips, J.B., 1948. Growth of the Sardine, *Sardinops caerulea* 1941-42 through 1946-47. Sacramento, CA: State of California Department of Natural Resources, Division of Fish and Game.

153. Prasad, R.R., 1948. *Life History of Clevelandia ios (Jordan and Gilbert)*. Stanford, CA: Stanford University.

154. Prescott, J.H., 1961. Unpublished data. Marineland of the Pacific, California.

155. Prince, E.D., D.W. Lee, C.A. Wilson, and J.M. Dean, 1986. Longevity and age validation of a tag-recaptured Atlantic sailfish, *Istiophorus platypterus* using dorsal spines and otoliths. *U.S. Fish and Wildlife Service Fishery Bulletin* 84: 493-502.

156. Probst, R.T. and E.L. Cooper, 1954. Age, growth, and production of the lake sturgeon (*Acipenser fulvescens*) in the Lake Winnebago region, Wisconsin. *Transactions of the American Fisheries Society* 84: 207-227.

157. Purkett, C.A.J., 1963. The paddlefish fishery of the Osage River and the Lake of the Ozarks, Missouri. *Transactions American Fisheries Society* 92: 239.

158. Pycha, R.L., 1956. Progress report on white sturgeon studies. *California Fish and Game* 42: 23-35.

159. Qasim, S.Q., 1957. The biology of *Blennius pholis* L. (Teleostei), in *Proceedings of the Zoological Society of London* 128: 161-207.

160. Randall, J.E., 1977. Contribution to the biology of the whitetip reef shark (*Triaenodon obesus*). *Pacific Science* 31: 143-164.

161. Raney, E.C., 1942. *Transactions American Fisheries Society* 71: 215.

162. Rawson, D.S., 1951. Studies of the fish of Great Slave Lake. *Journal of Fisheries Research Board Canada* 8: 207.

163. Reed, R.J. and A.D. McCall, 1988. Changing the size limit: how it could affect California halibut fisheries. *California Cooperative Oceanology Fisheries Investigative Report* 29: 158-166.

164. Reighard, J., 1915. An ecological reconnaissance of the fishes of Douglas Lake, Cheboygon County, Michigan, in midsummer. *U. S. Fish Commission Bulletin* 33: 215.

165. Reimchen, T.E., 1992. Extended longevity in a large-bodied stickleback, *Gasterosteus*, population. *Canadian Field Naturalist* 106: 129-131.

166. Reimers, N., 1979. A history of a stunted brook trout population in an Alpine lake: a lifespan of 24 years. *California Fish and Game* 65: 196-215.

167. Roach, L.S., 1948. *Ohio Conservation Bulletin* 12: 12.

168. Runnstrom, S., 1936. A study of the life history and migrations of the Norwegian spring-herring, based on the analysis of the winter rings and summer zones of the scale. *Fiskeridirektoratets skrifter Serie Havundersokelser* 5.

169. Sandeman, E.J., 1969. Age determination & growth rate of redfish, *Sebastes spp.*, from selected areas around Newfoundland. *Research Bulletin International Commision for the Northwest Atlantic Fisheries* 6: 79-106.

170. Sato, K., 1952. *Japanese Angler*. Tokyo: Foreign Affairs Association of Japan.

171. Schaeffer, M.B. and J.C. Marr, 1948. *U. S. Bureau Fisheries Bulletin* 51: 187.

172. Schmidt, U., 1955. Beitrage zur biologie des Kohles (*Gadus virens*) in den Isandisolen Gewassen. *Ber. deut. wiss. Komm. Meeresforsch* 14: 46.

173. Schneider, J.C., P.H. Eschmeyer, and W.R. Crowe, 1977. Longevity, survival and harvest of tagged walleyes in Lake Gogebic, Michigan. *Transactions of the American Fisheries Society* 106: 566-568.

174. Schultz, L.P., 1933. The age and growth of *Atherinops affinis oregonia* Jordan and Snyder and other subspecies of baysmelt along the Pacific Coast of the United States. *Washington State University Publication, Biology* 2: 45-102.

175. Schwartz, F.J. and R. Jachowski, 1965. [No Title]. *Chesapeake Science* 6: 226.

176. Sciences, M.A.E., 1987. Ecology of Important Fisheries Species Offshore California. Washington, D. C.: Min. Managment Service, U. S. Department Interior.

177. Scott, W.B. and E.J. Crossman, 1973. Freshwater Fishes of Canada. *Fisheries Research Board Canada* Bulletin 84: 966.

178. Seemann, W., 1961. Der Sehlendorfer Binnensee...VI. Die Fisherei im Sehlendorfer Binnensee. *Fisch. Hilfswiss* 9: 603.

179. Shields, J.T., 1958. Fish and Parks Mimeo. Report. South Dakota Department Game.

180. Shimada, B.M. and W.G. vanCampen, 1950. Morphometry, growth and age of tunas. *U. S. Fish Wildlife Service Special Scientific Report on Fisheries* 22.

181. Sigler, W.F., 1949. Life History of the White Bass in Storm Lake, Iowa. *Iowa Agricultural Experiment Station Research Bulletin* 366: 203.

182. Silliman, R.P., 1943. *Studies on the Pacific pilchard or sardine (Sardinops caerulea) 5. Method of computing mortalities and replacements.* Chicago: U. S Fish and Wildlife Service Special Report No 24.

183. Silliman, R.P., 1950. A method of computing mortalities and replacements. *U. S. Fish Wildlife Service Special Scientific Report Fisheries* 15: 168.

184. Sivalingam, S., 1956. Age and growth of *Taeniotoca lateralis. Ceylon Journal of Science* 7: 135.

185. Smith, W.E. and R.W. Saalfeld, 1955. Studies on Columbia River smelt, *Thaleichthys pacificus* (Richardson). *Washington Department of Fishery Research Papers* 1: 3-26.

186. Sprules, W.M., 1952. The Arctic char of the west coast of Hudson Bay. *Journal of Fisheries Research Board Canada* 9: 1.

187. Strawn, K., 1958. Life history of the pygmy seahorse, *Hippocampus zosterae* (Jordan and Gilbert) at Cedar Key, Florida. *Copeia* : 16-22.

188. Stroud, R.H., 1948. Growth of the basses and black crappie in Norris Reservoir, Tennessee. *Journal of Tennessee Academy of Science* 23: 31.

189. Stroud, R.H., 1955. *Fisheries Report*. Boston: Massachusetts Division of Fish and Game.

190. Summer, F.H., 1962. Migration and growth of coastal cutthroat trout in Tillamook County, Oregon. *Transactions American Fisheries Society* 91: 71-83.

191. Taft, A.C. and G.I. Murphy, 1950. Life history of the Sacramento squawfish (*Atychocheilus grandis*). *California Fish and Game* 36: 147-164.

192. Takai, T., 1959. Studies on the morphology, ecology, and culture of the important apodal fishes, *Muranesox cinereus* (Forskal) and *Canger myriaster* (Brevoort). *Journal Shimonoseki Con. Fisheries* 8: 209.

193. Tanaka, S., C.T. Chen, and K. Mizue, 1978. Studies on sharks: 16. Age and growth of Eiraku shark *Galeorhinus japonicus*. *Bulletin Nagasake University* 45: 19-28.

194. Tester, A.L., 1937. Populations of herring (*Clupea pallasi*) in the coastal waters of British Columbia. *Journal of the Biological Board of Canada* 3: 108-144.

195. Turner, C.H., E.E. Ebert, and R.R. Given, 1969. Manmade reef ecology. *California Fish and Game* 146: 221.

196. vanOosten, J., 1937. The age, growth, and sex ratio of the Lake Superior longjaw, *Leucichthys zenithicus* (Jordan and Evermann). *Papers Michigan Academy Science* 22: 691.

197. Walford, L.A., 1932. The California barracuda (*Sphyraena argentea*) I. Life history of the California barracuda. II. A bibliography of barracudas (Sphyraenidae). *California Division of Fish and Game Bulletin* 37: 1.

198. Wang, J.C.S., 1986. Fishes of the Sacramento-San Joaquin estuary and adjacent waters, California: A guide to the early life histories. Sacramento: California Department of Water Resources.

199. Washington, P., 1970. Occurrence on the high seas of a steelhead trout in its ninth year. *California Fish and Game* 56: 312-314.

200. Wigley, R.L., 1959. Life history of the sea lamprey of Cayuga Lake (N.Y.). *U. S. Fish Wildlife Service Fishery Bulletin* 59: 561.

201. Wohlschlag, D.E., 1954. Growth pecularities of the cisco, *Coregonus sardinella* (Valenciennes), in the vicinity of Point Barrow, Alaska. *Stanford Ichthyological Bulletin* 4: 189-209.

202. Wohlschlag, D.E., 1954. Mortality rates of whitefish in an Arctic lake. *Ecology* 35: 388-396.

203. Wydoski, R.S. and D.E. Bennett, 1973. Contributions to the life history of the silver surfperch (*Hyperprosopon ellipticum*) from the Oregon coast. *California Fish and Game* 59: 178-190.

204. Wydoski, R.S. and R.R. Whitney, 1979. *Inland Fishes of Washington*. Seattle, WA: University Washington Press.

205. Yapchiongo, J.V., 1949. *Hypomesus pretiosus*: Its development and early life history. *National Applied Science Bulletin* 9: 3-108.

206. Young, P.H., 1963. The kelp bass (*Paralabrax clathratus*) and its fishery, 1947-1958. *California Fish and Game* 122: 1-67.

Index of Scientific Binomials

A

B

C

D

E

F

G

H

I

J

K

L

M

N

O

P

Q

R

S

T

U

V

W

X

Z

Index of Common Names

A

B

C

D

E

F

G

H

I

J

K

L

M

N

O

P

Q

R

S

T

U

V

W

Y

Z

In the series Odense Monographs on Population Aging the following volumes have been published:

MONOGRAPHS ON POPULATION AGING

General Editors
Bernard Jeune and James W. Vaupel

Vol. 1
Development of Oldest-Old Mortality, 1950-1990:
Evidence from 28 Developed Countries
Väinö Kannisto

Vol. 2
Exceptional Longevity:
From Prehistory to the Present
Bernard Jeune and James W. Vaupel (Eds.)

Vol. 3
The Advancing Frontier of Survival:
Life Tables for Old Age
Väinö Kannisto

Vol. 4
Population Data at a Glance:
Shaded Contour Maps of Demographic Surfaces over Age and Time
James W. Vaupel, Zhenglian Wang, Kirill F. Andreev,
and Anatoli I. Yashin

Vol. 5

The Force of Mortality at Ages 80 to 120

A.R. Thatcher, V. Kannnisto, and J.W. Vaupel

Vol. 6

Validation of Exceptional Longevity

Bernard Jeune and James W. Vaupel (Eds.)

Vol. 7

Mechanisms of Aging and Mortality: The Search for New Paradigms

Kenneth G. Manton and Anatoli I. Yashin

Vol. 8

Longevity Records: Life spans of Mammals, Birds, Amphibians, Reptiles, and Fish

James R. Carey and Debra S. Judge

Aging Research Center

Odense University